TIME TRAVEL

AND OTHER MATHEMATICAL BEWILDERMENTS

TIME TRAVEL

AND OTHER MATHEMATICAL BEWILDERMENTS

.
.
.

MARTIN GARDNER

W. H. FREEMAN AND COMPANY   NEW YORK

Library of Congress Cataloguing-in-Publication Data

Gardner, Martin, 1914–
    Time travel and other mathematical bewilderments.

    Includes index.
    1. Mathematical recreations.   I. Title.
QA95.G325   1987   793.7′4   87-11849
ISBN 0-7167-1924-X
ISBN 0-7167-1925-8 (pbk.)

Printed in the United States of America

2 3 4 5 6 7 8 9 0 VB 6 5 4 3 2 1 0 8 9 8

To David A. Klarner

for his many splendid contributions
to recreational mathematics,
for his friendship over the years,
and in gratitude for many other things.

CONTENTS

PREFACE **xi**

CHAPTER ONE
Time Travel   **1**
•
•

CHAPTER TWO
Hexes and Stars   **15**
•
•

CHAPTER THREE
Tangrams, Part 1   **27**
•
•

CHAPTER FOUR
Tangrams, Part 2   **39**
•
•

CHAPTER FIVE
Nontransitive Paradoxes   **55**
•
•

CHAPTER SIX
Combinatorial Card Problems   **71**
•
•

CHAPTER SEVEN
Melody-Making Machines   **85**
•
•

CHAPTER EIGHT
Anamorphic Art   **97**
•
•

CHAPTER NINE
The Rubber Rope and Other Problems   **111**
•
•

CHAPTER TEN
Six Sensational Discoveries   125
.
.
.

CHAPTER ELEVEN
The Császár Polyhedron   139
.
.
.

CHAPTER TWELVE
Dodgem and Other Simple Games   153
.
.
.

CHAPTER THIRTEEN
Tiling with Convex Polygons   163
.
.
.

CHAPTER FOURTEEN
Tiling with Polyominoes, Polyiamonds, and Polyhexes   177
.
.
.

CHAPTER FIFTEEN
Curious Maps   189
.
.
.

CHAPTER SIXTEEN
The Sixth Symbol and Other Problems   205
.
.
.

CHAPTER SEVENTEEN
Magic Squares and Cubes   213
.
.
.

CHAPTER EIGHTEEN
Block Packing   227
.
.
.

CHAPTER NINETEEN
Induction and Probability   241
.
.
.

CHAPTER TWENTY
Catalan Numbers   **253**
•
•
•

CHAPTER TWENTY-ONE
Fun with a Pocket Calculator   **267**
•
•
•

CHAPTER TWENTY-TWO
Tree-Plant Problems   **227**
•
•
•

INDEX OF NAMES   **291**

P R E F A C E

Herewith the twelfth collection of my columns from *Scientific American*. As usual, they have been corrected, updated, and expanded, mostly on the basis of letters from knowledgeable and alert readers.

MG

TIME TRAVEL
AND OTHER MATHEMATICAL BEWILDERMENTS

ONE

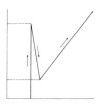

.
.
.

Time Travel

"It's against reason," said Filby.
"What reason?" said the Time
Traveller.

—H.G. WELLS,
The Time Machine

H. G. Wells's short novel *The Time Machine,* an undisputed masterpiece of science fiction, was not the first story about a time machine. That distinction belongs to "The Clock That Went Backward," a pioneering but mediocre yarn by Edward Page Mitchell, an editor of the New York *Sun.* It was published anonymously in the *Sun* on September 18, 1881, seven years before young Wells (he was only 22) wrote the first version of his famous story.

Mitchell's tale was so quickly forgotten that science-fiction buffs did not even know of its existence until Sam Moskowitz reprinted it in his anthology of Mitchell's stories, *The Crystal Man* (1973). Nor did anyone pay much attention to Wells's fantasy when it was serialized in 1888 in *The Science*

Schools Journal under the horrendous title "The Chronic Argonauts," Wells himself was so ashamed of this clumsily written tale that he broke it off after three installments and later destroyed all the copies he could find. A completely rewritten version, "The Time Traveller's Story," was serialized in *The New Review* beginning in 1894. When it came out as a book in 1895, it brought Wells instant recognition.

One of the many remarkable aspects of Wells's novella is the introduction in which the Time Traveller (his name is not revealed, but in Wells's first version he is called Dr. Nebo-gipfel) explains the theory behind his invention. Time is a fourth dimension. An instantaneous cube cannot exist. The cube we see is at each instant a cross section of a "fixed and unalterable" four-dimensional cube having length, breadth, thickness, and duration. "There is no difference between Time and any of the three dimensions of Space," says the Time Traveller, "except that our consciousness moves along it." If we could view a person from outside our space-time (the way human history is viewed by the Eternals in Isaac Asimov's *The End of Eternity* or by the Tralfamadorians in Kurt Vonnegut's *Slaughterhouse-Five*), we would see that person's past, present, and future all at once, just as in 3-space we see all parts of a wavy line that traces on a time chart the one-dimensional spatial movements of mercury in a barometer.

Reading these remarks today, one might suppose that Wells had been familiar with Hermann Minkowski's great work of tidying up Einstein's special theory of relativity. The line along which our consciousness crawls is, of course, our "world line": the line that traces our movements in 3-space on a four-dimensional Minkowski space-time graph. (*My World Line* is the title of George Gamow's autobiography.) But Wells's story appeared in its final form ten years before Einstein published his first paper on relativity!

When Wells wrote his story, he regarded the Time Traveller's theories as little more than metaphysical hanky-panky designed to make his fantasy more plausible. A few decades later physicists were taking such hanky-panky with the utmost seriousness. The notion of an absolute cosmic time, with absolute simultaneity between distant events, was swept out of physics by Einstein's equations. Virtually all physicists now agree that if an astronaut were to travel to a distant star and back, moving at a velocity close to that of light, he could in theory travel thousands of years into the earth's future. Kurt Gödel constructed a rotating cosmological model in which one can, in principle, travel to any point in the world's past as well as future, although travel to the past is ruled out as physically impossible. In 1965 Richard P. Feynman received a Nobel prize for his space-time approach to quantum mechanics in which antiparticles are viewed as particles momentarily moving into the past.

Hundreds of science-fiction stories have been written about time travel, many of them raising questions about time and causality that are as profound as they are sometimes funny. To give the most hackneyed example, suppose you traveled back to last month and shot yourself through the head. Not only do you know before making the trip that nothing like this happened but, assuming that somehow you could murder your earlier self, how could you exist next month to make the trip? Fredric Brown's "First Time Machine" opens with Dr. Grainger exhibiting his machine to three friends. One of them uses the device to go back sixty years and kill his hated grandfather when the man was a youth. The story ends sixty years later with Dr. Grainger showing his time machine to two friends.

It must not be thought that logical contradictions arise only when people travel in time. The transportation of anything can lead to paradox. There is a hint of this in Wells's story. When the Time Traveller sends a small model of his machine into the past or the future (he does not know which), his guests raise two objections. If the time machine went into the future, why do they not see it now, moving along its world line? If it went into the past, why did they not see it there before the Time Traveller brought it into the room?

One of the guests suggests that perhaps the model moves so fast in time it becomes invisible, like the spokes of a rotating wheel. But what if a time-traveling object stops moving? If you have no memory of a cube on the table Monday, how could you send it back to Monday's table on Tuesday? And if on Tuesday you go into the future, put the cube on the table Wednesday, then return to Tuesday, what happens on Wednesday if on Tuesday you destroy the cube?

Objects carried back and forth in time are sources of endless confusion in certain science-fiction tales. Sam Mines once summarized the plot of his own story, "Find the Sculptor," as follows: "A scientist builds a time machine, goes 500 years into the future. He finds a statue of himself commemorating the first time traveler. He brings it back to his own time and it is subsequently set up in his honor. You see the catch here? It had to be set up in his own time so that it would be there waiting for him when he went into the future to find it. He had to go into the future to bring it back so it could be set up in his own time. Somewhere a piece of the cycle is missing. When was the statue made?"

A splendid example of how paradox arises, even when nothing more than messages go back in time, is provided by the conjecture that tachyons, particles moving faster than light, might actually exist. Relativity theory leaves no escape from the fact that anything moving faster than light would move backward in time. This is what inspired A. H. Reginald Buller, a Canadian botanist, to write his often quoted limerick:

There was a young lady named Bright
Who traveled much faster than light.
She started one day
In the relative way,
And returned on the previous night.

Tachyons, if they exist, clearly cannot be used for communication. G. A. Benford, D. L. Book, and W. A. Newcomb (of "Newcomb's paradox," the topic of two chapters in my *Knotted Doughnuts and Other Mathematical Entertainments,* W. H. Freeman and Company, 1986), have chided physicists who are searching for tachyons for overlooking this. In "The Tachyonic Antitelephone," they point out that certain methods of looking for tachyons are based on interactions that make possible, in theory, communication by tachyons. Suppose physicist Jones on the earth is in communication by tachyonic antitelephones with physicist Alpha in another galaxy. They make the following agreement. When Alpha receives a message from Jones, he will reply immediately. Jones promises to send a message to Alpha at three o'clock earth time, if and only if he has not received a message from Alpha by one o'clock. Do you see the difficulty? Both messages go back in time. If Jones sends his message at three, Alpha's reply could reach him before one. "Then," as the authors put it, "the exchange of messages will take place if and only if it does not take place . . . a genuine . . . causal contradiction." Large sums of money have already gone down the drain, the authors believe, in efforts to detect tachyons by methods that imply tachyonic communication and are therefore doomed to failure.

Time dilation in relativity theory, time travel in Gödel's cosmos, and reversed time in Feynman's way of viewing antiparticles are so carefully hedged by other laws that contradictions cannot arise. In most time-travel stories the paradoxes are skirted by leaving out any incident that would generate a paradox. In some stories, however, logical contradictions explicitly arise. When they do, the author may leave them paradoxical to bend the reader's mind or may try to escape from paradox by making clever assumptions.

Before discussing ways of avoiding the paradoxes, brief mention should be made of what might be called pseudo-time-travel stories in which there is no possibility of contradiction. There can be no paradox, for example, if one simply observes the past but does not interact with it. The electronic machine in Eric Temple Bell's "Before the Dawn," which extracts motion pictures of the past from imprints left by light on ancient rocks, is as free of possible paradox as watching a video tape of an old television show. And paradox

cannot arise if a person travels into the future by going into suspended anima-
tion, like Rip van Winkle, or Woody Allen in his motion picture *Sleeper,* or
the sleepers in such novels as Edward Bellamy's *Looking Backward* or Wells's
When the Sleeper Wakes. No paradox can arise if one dreams of the past (as in
Mark Twain's *A Connecticut Yankee at King Arthur's Court,* or in the 1986
motion picture *Peggy Sue Got Married*), or goes forward in a reincarnation, or
lives for a while in a galaxy where change is so slow in relation to earth time
that when he returns, centuries on the earth have gone by. But when someone
actually travels to the past or the future, interacts with it and returns, enormous difficulties arise.

In certain restricted situations paradox can be avoided by invoking Minkowski's "block universe," in which all history is frozen, as it were, by one
monstrous space-time graph on which all world lines are eternal and unalterable. From this deterministic point of view one can allow certain kinds of time
travel in either direction, although one must pay a heavy price for it. Hans
Reichenbach, in a muddled discussion in *The Philosophy of Space and Time*
(Dover, 1957, pp. 140 – 142), puts it this way: Is it possible for a person's world
line to "loop" in the sense that it returns him to a spot in space-time, a spot
very close to where he once had been and where some kind of interaction, such
as speech, occurs between the two meeting selves? Reichenbach argues that
this cannot be ruled out on logical grounds; it can only be ruled out on the
ground that we would have to give up two axioms that are strongly confirmed
by experience: (1) A person is a unique individual who maintains his identity
as he ages, and (2) a person's world line is linearly ordered so that what he
considers "now" is always a unique spot along the line. (Reichenbach does not
mention it, but we would also have to abandon any notion of free will.) If we
are willing to give up these things, says Reichenbach, we can imagine without
paradox certain kinds of loops in a person's world line.

Reichenbach's example of a consistent loop is as follows. One day you meet
a man who looks exactly like you but who is older. He tells you he is your older
self who has traveled back in time. You think him insane and walk on. Years
later you discover how to go back in time. You visit your younger self. You are
compelled to tell him exactly what your older duplicate had told you when
you were younger. Of course, he thinks that you are insane. You separate.
Each of you leads a normal life until the day comes when your younger self
makes the trip back in time.

Hilary Putnam, in "It Ain't Necessarily So," argues in similar fashion that
such world-line loops need not be contradictory. He draws a Feynman graph
(*see* Figure 1) on which particle pair-production and pair-annihilation are
replaced by person pair-production and pair-annihilation. The zigzag line is

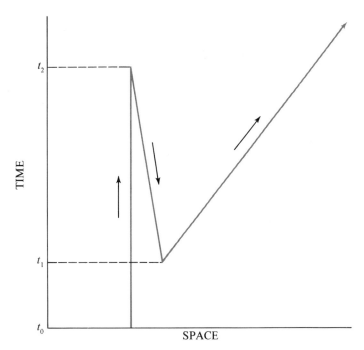

Figure 1 Feynman graph for a time traveler to the past

the world line of time traveler Smith. At time t_2 he goes back to t_1, converses with his younger self, then continues to lead a normal life. How would this be observed by someone whose world line is normal? Simply put a ruler at the bottom of the chart, its edge parallel to the space axis, and move it slowly upward. At t_0 you see young Smith. At t_1 an older Smith suddenly materializes out of thin air in the same room along with an anti-Smith, who is seated in his time machine and living backward. (If he is smoking, you see his cigarette butt lengthen into a whole cigarette, and so on.) Perhaps the two forward Smiths converse. Finally, at t_2, young Smith, backward Smith, and the backward-moving time machine vanish. The older Smith and his older time machine continue on their way. The fact that we can draw a space-time diagram of these events, Putnam insists, is proof that they are logically consistent.

It is true that they are consistent, but note that Putnam's scenario, like Reichenbach's, involves such weak interaction between the Smiths that it evades the deeper contradictions that arise in time-travel fiction. What happens if the older Smith kills the younger Smith? Will Putnam kindly supply a Feynman graph?

There is only one good way out, and science-fiction scribblers have been using it for more than half a century. According to Sam Moskowitz, the device was first explicitly employed to resolve time-travel paradoxes by David R. Daniels in "Branches of Time," a tale that appeared in *Wonder Stories* in 1934. The basic idea is as simple as it is fantastic. Persons can travel to any point in the future of their universe, with no complications, but the moment they enter the past, the universe splits into two parallel worlds, each with its own time track. Along one track rolls the world as if no looping had occurred. Along the other track spins the newly created universe, its history permanently altered. When I say "newly created," I speak, of course, from the standpoint of the time traveler's consciousness. For an observer in, say, a fifth dimension the traveler's world line simply switches from one space-time continuum to another on a graph that depicts all the universes branching like a tree in a metauniverse.

Forking time paths appear in many plays, novels, and short stories by non-science-fiction writers. J. B. Priestley uses it in his popular play *Dangerous Corner,* as Lord Dunsany had done earlier in his play *If.* Mark Twain discusses it in *The Mysterious Stranger.* Jorge Luis Borges plays with it in his "Garden of Forking Paths." But it was the science-fiction writers who sharpened and elaborated the concept.

Let us see how it works. Suppose you go back to the time of Napoleon in Universe 1 and assassinate him. The world forks. You are now in Universe 2. If you like, you can return to the present of Universe 2, a universe in which Napoleon had been mysteriously murdered. How much would this world differ from the old one? Would you find a duplicate of yourself there? Maybe. Maybe not. Some stories assume that the slightest alteration of the past would introduce new causal chains that would have a multiplying effect and produce vast historical changes. Other tales assume that history is dominated by such powerful overall forces that even major alterations of the past would damp out and the future would soon be very much the same.

In Ray Bradbury's "A Sound of Thunder," Eckels travels back to an ancient geological epoch under elaborate precautions to prevent any serious alteration of the past. For example, he wears an oxygen mask to prevent his microbes from contaminating animal life. But Eckels violates a prohibition and accidentally steps on a living butterfly. When he returns to the present, he notices subtle changes in the office of the firm that arranged his trip. He is killed for having illegally altered the future.

Hundreds of other stories by fantasy and science-fiction writers have played variations on this theme. One of the saddest is Lord Dunsany's "Lost" (in *The Fourth Book of Jorkens,* 1948). A man travels to his past, by way of an Oriental charm, to right some old mistakes. Of course, this alters history. When he gets back to the present, he is missing his wife and home. "Lost! Lost!" he cries. "Don't go back down the years trying to alter anything. Don't even wish to . . . . And, mind you, the whole length of the Milky Way is more easily traveled than time, amongst whose terrible ages I am lost."

It is easy to see that in such a metacosmos of branching time paths, it is not possible to generate paradox. The future is no problem. If you travel to next week, you merely vanish for a week and reappear in the future a week younger than you would have been. But if you go back and murder yourself in your crib, the universe obligingly splits. Universe 1 goes on as before, with you vanishing from it when you grow up and make the trip back. Perhaps this happens repeatedly, each cycle creating two new worlds. Perhaps it happens only once. Who knows? In any case, Universe 2 with you and the dead baby in it rolls on. You are not annihilated by your deed, because now you are an alien from Universe 1 living in Universe 2.

In such a metacosmos it is easy (as many science-fiction writers have done) to fabricate duplicates of yourself. You can go back a year in Universe 1, live for a year with yourself in Universe 2, then again go back a year to visit two replicas of yourself in Universe 3. Clearly, by repeating such loops you can create as many replicas of yourself as you please. They are genuine replicas, not pseudo-replicas as in the scenarios by Reichenbach and Putnam. Each has his independent world line. History might become extremely chaotic, but there is one type of event that can never occur: a logically contradictory one.

This vision of a metacosmos containing branching worlds may seem crazy, but respectable physicists have taken it quite seriously. In Hugh Everett III's Ph.D. thesis " 'Relative State' Formulation of Quantum Mechanics" (*Reviews of Modern Physics* 29, July 1957, pp. 454–462) he outlines a metatheory in which the universe at every micromicroinstant branches into countless parallel worlds, each a possible combination of microevents that could occur as a result of microlevel uncertainty. The paper is followed by John A. Wheeler's favorable assessment in which he points out that classical physicists were almost as uncomfortable at first with the radical notions of general relativity.

"If there are infinite universes," wrote Fredric Brown in *What Mad Universe,* "then all possible combinations must exist. Then, somewhere, *everything must be true* . . . . There is a universe in which Huckleberry Finn is a real person, doing the exact things Mark Twain described him as doing. There are, in fact, an infinite number of universes in which a Huckleberry Finn is doing every possible variation of what Mark Twain *might* have described him as doing . . . . And infinite universes in which the states of existence are such that we would have no words or thoughts to describe them or to imagine them."

What if the universe never forks? Suppose there is only one world, this one, in which all world lines are linearly ordered and objects preserve their identity, come what may. Brown considers this possibility in his story "Experiment." Professor Johnson holds a brass cube in his hand. It is six minutes to three o'clock. At exactly three, he tells his colleagues, he will place the cube on his time machine's platform and send it five minutes into the past.

"Therefore," he remarks, "the cube should, at five minutes before three, vanish from my hand and appear on the platform, five minutes before I place it there."

"How can you place it there, then?" asked one of his colleagues.

"It will, as my hand approaches, vanish from the platform and appear in my hand to be placed there."

At five minutes to three the cube vanishes from Professor Johnson's hand and appears on the platform, having been sent back five minutes in time by his future action of placing the cube on the platform at three.

"See? Five minutes before I shall place it there, it *is* there!"

"But," says a frowning colleague, "what if, now that it has already appeared five minutes before you place it there, you should change your mind about doing so and *not* place it there at three o'clock? Wouldn't there be a paradox of some sort involved?"

Professor Johnson thinks this is an interesting idea. To see what happens, he does not put the cube on the platform at three.

There is no paradox. The cube remains. But the entire universe, including Professor Johnson, his colleagues, and the time machine, disappears.

ADDENDUM

J. A. Lindon, a British writer of comic verse, sent me his sequel to the limerick about Miss Bright:

> When they questioned her, answered Miss Bright,
> "I was there when I got home that night;
>   So I slept with myself,
>   Like two shoes on a shelf,
> Put-up relatives shouldn't be tight!"

Ned Block wrote to say he had heard the following blue version from a student at M.I.T.:

There was a young couple named Bright
Who could make love much faster than light.
They started one day
In the relative way,
And came on the previous night.

Many readers called attention to two difficulties that could arise from time travel in either direction. If travelers stay at the same spot in space-time, relative to the universe, the earth would no longer be where it was. They might find themselves in empty space, or inside something solid. In the latter case, would the solid body prevent them from arriving? Would one or the other be shoved aside? Would there be an explosion?

The second difficulty is thermodynamic. After the time traveler departs, the universe will have lost a bit of mass-energy. When he arrives, the universe gains back the same amount. During the interval between leaving and arriving, the universe would seem to be violating the law of mass-energy conservation.

I mentioned briefly what is now called the "many-worlds interpretation" of QM (quantum mechanics). The best reference is a 1973 collection of papers on the topic, edited by Bryce DeWitt and Neill Graham. Assuming that the universe constantly splits into billions of parallel worlds, the interpretation provides an escape from the indeterminism of the Copenhagen interpretation of QM, as well as from the many paradoxes that plague it.

Some physicists who favor the many-worlds interpretation have argued that the countless duplicate selves and parallel worlds produced by the forking paths are not "real," but only artifacts of the theory. In this interpretation of the many-worlds interpretation, the theory collapses into no more than a bizarre way of saying the same things that are said in the Copenhagen interpretation. Everett himself, in his original 1957 thesis, added in proof this famous footnote:

> In reply to a preprint of this article some correspondents have raised the question of the "transition from possible to actual," arguing that in "reality" there is—as our experience testifies—no such splitting of observer states, so that only one branch can ever actually exist. Since this point may occur to other readers the following is offered in explanation.
>
> The whole issue of the transition from "possible" to "actual" is taken care of in the theory in a very simple way—there is no such transition, nor is such a transition necessary for the theory to be in

accord with our experience. From the viewpoint of the theory *all* elements of a superposition (all "branches") are "actual," none any more "real" than the rest. It is unnecessary to suppose that all but one are somehow destroyed, since all the separate elements of a superposition individually obey the wave equation with complete indifference to the presence or absence ("actuality" or not) of any other elements. This total lack of effect of one branch on another also implies that no observer will ever be aware of any "splitting" process.

Arguments that the world picture presented by this theory is contradicted by experience, because we are unaware of any branching process, are like the criticism of the Copernican theory that the mobility of the earth as a real physical fact is incompatible with the common sense interpretation of nature because we feel no such motion. In both cases the argument fails when it is shown that the theory itself predicts that our experience will be what it in fact is. (In the Copernican case the addition of Newtonian physics was required to be able to show that the earth's inhabitants would be unaware of any motion of the earth.)

The many-worlds interpretation has been called a beautiful theory nobody can believe. Nevertheless, a number of top physicists have indeed accepted — some still do — its outrageous multiplicity of logically possible worlds. Here is DeWitt defending it in "Quantum Mechanics and Reality," a 1970 article reprinted in the collection he edited with Graham:

> The obstacle to taking such a lofty view of things, of course, is that it forces us to believe in the reality of all the simultaneous worlds . . . in each of which the measurement has yielded a different outcome. Nevertheless, this is precisely what [the inventors of the theory] would have us believe . . . . This universe is constantly splitting into a stupendous number of branches, all resulting from the measurementlike interactions between its myriads of components. Moreover, every quantum transition taking place on every star, in every galaxy, in every remote corner of the universe is splitting our local world on earth into myriads of copies of itself.
>
> I still recall vividly the shock I experienced on first encountering this multiworld concept. The idea of 10^{100+} slightly imperfect copies of oneself all constantly splitting into further copies, which

ultimately become unrecognizable, is not easy to reconcile with commonsense.

Although John Wheeler originally supported the many-worlds interpretation, he has since abandoned it. I quote from the first chapter of his *Frontiers of Time* (Center for Theoretical Physics, 1978):

> Imaginative Everett's thesis is, and instructive, we agree. We once subscribed to it. In retrospect, however, it looks like the wrong track. First, this formulation of quantum mechanics denigrates the quantum. It denies from the start that the quantum character of nature is any clue to the plan of physics. Take this Hamiltonian for the world, that Hamiltonian, or any other Hamiltonian, this formulation says. I am a principle too lordly to care which, or why there should be any Hamiltonian at all. You give me whatever world you please, and in return I give you back many worlds. Don't look to me for help in understanding this universe.
>
> Second, its infinitely many unobservable worlds make a heavy load of metaphysical baggage. They would seem to defy Mendeléev's demand of any proper scientific theory, that it should "expose itself to destruction."
>
> Wigner, Weizsäcker, and Wheeler have made objections in more detail, but also in quite contrasting terms, to the relative-state or many-worlds interpretation of quantum mechanics. It is hard to name anyone who conceives of it as a way to uphold determinism.

In the paper titled "Rotating Cylinders and the Possibility of Global Causality Violation," physicist Frank Tipler raised the theoretical possibility of constructing a machine that would enable one to go forward or backward in time. (Tipler is one of the few remaining enthusiasts for the many-worlds interpretation, and the coauthor of a controversial book, *The Anthropic Cosmological Principle,* Oxford University Press, 1986). Taking off from Gödel's rotating cosmos and from recent work on the space-time pathologies surrounding black holes, Tipler imagines a massive cylinder, infinitely long, and rotating so rapidly that its surface moves faster than half the speed of light. Space-time near the cylinder would be so distorted that, according to Tipler's calculations, astronauts could orbit the cylinder, going with or against its spin, and travel into their past or future.

Tipler speculated on the possibility that such a machine could be built with a cylinder of finite length and mass, but later concluded that such a device was impossible to construct with any known forms of matter and force. Such doubts did not inhibit Poul Anderson from using Tipler's cylinder for time travel in his novel *The Avatar,* nor did it stop Robert Forward from writing "How to Build a Time Machine" (*Omni,* May 1980). "We already know the theory," *Omni* editors commented above Forward's backward article, "All that's needed is some advanced engineering."

I close with two pearls of wisdom from the stand-up comic "Professor" Irwin Corey: "The past is behind us, and the future lies ahead."

B I B L I O G R A P H Y

Travelers in Time. Philip Van Doren Stern, ed. Doubleday, 1947.

Science Fiction Adventures in Dimension. Groff Conklin, ed. Vanguard, 1953.

"It Ain't Necessarily So." Hilary Putnam, in *The Journal of Philosophy* 59, October 25, 1962, pp. 658–671; reprinted in Putnam's *Philosophical Papers,* Vol. 1, Cambridge University Press, 1975.

"Is Time Travel Possible?" J. J. C. Smart, in *The Journal of Philosophy* 60, 1963, pp. 237–241.

"Time and Fiction in Drama." J. B. Priestley, in *Man and Time.* Doubleday, 1964.

"Measuring Measuring Rods." John C. Graves and James E. Roper, in *Philosophy of Science* 32, January 1965, pp. 39–56.

"On Going Backward in Time." John Earman, in *Philosophy of Science* 34, September 1967, pp. 211–222.

"Particles That Go Faster Than Light." Gerald Feinberg, in *Scientific American,* February 1970, pp. 69–77.

"The Tachyonic Antitelephone." G. A. Benford, D. L. Bock, and W. A. Newcomb, in *Physical Review D* 2, July 15, 1970, pp. 263–265.

Time and the Space-Traveler. L. Marder. Allen & Unwin, 1971.

"Tachyon Paradoxes." L. S. Schulman, in *American Journal of Physics* 39, May 1971, pp. 481–484.

The Many-Worlds Interpretation of Quantum Mechanics. Bryce S. DeWitt and Neill Graham, eds. Princeton University Press, 1973.

"Rotating Cylinders and the Possibility of Global Causality Violation." Frank J. Tipler, in *Physical Review D* 9, April 15, 1974, pp. 2203–2206.

"The Paradoxes of Time Travel." David Lewis, in *American Philosophical Quarterly* 13, 1976, pp. 145–152.

"Time and the Nth Dimension" and "Lost and Parallel Worlds." *The Visual Encyclopedia of Science Fiction*. Brian Ash, ed. Harmony, 1977.

"Time Travel," "Time Paradoxes," "Alternate Worlds," and "Parallel Worlds." *The Science Fiction Encyclopedia*. Peter Nicholls, ed. Doubleday, 1979.

"Time Travel and Other Universes." Chapter 5 of *The Science in Science Fiction*. Peter Nicholls, ed. Knopf, 1983.

TWO

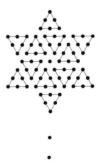

Hexes and Stars

Ancient Greek mathematicians, particularly the Pythagoreans, were entranced by figurate numbers: numbers that could be represented by arranging points in regular patterns on a plane or in space. Among the plane figurate numbers, the most studied were the polygonal numbers. Figure 2 shows how the first four polygonal numbers — triangular, square, pentagonal, and hexagonal — are built up as the partial sums of simple arithmetic progressions. Triangular numbers are partial sums of the counting numbers $1 + 2 + 3 + 4 + \ldots$ Square numbers are formed by successive addition of the consecutive odd numbers $1 + 3 + 5 + 7 + \ldots$ Pentagonal numbers derive from the progression $1 + 4 + 7 + 10 + \ldots$ and hexagonal numbers from the progression $1 + 5 + 9 + 13 + \ldots$ The respective differences are 1, 2, 3,. . . .

The study of figurate numbers belongs to a branch of number theory called Diophantine analysis, which has to do with finding integral solutions of equations. An enormous effort was made by the great pioneers of number theory in studying the properties of polygonal numbers. Most of this work is crisply summarized in the second volume of Leonard E. Dickson's *History of the Theory of Numbers* (Carnegie Institute, 1919).

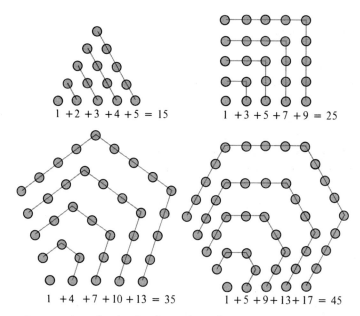

1 +2 +3 +4 +5 = 15

1 +3 +5 +7 +9 = 25

1 +4 +7 +10 +13 = 35

1 +5 +9+ 13+ 17 = 45

Figure 2 Construction of order-5 polygonal numbers

Let us start with a classic problem that was solved by Leonhard Euler in 1730. How can we find all the numbers that are both square and triangular? The formula for the nth triangular number is $\frac{1}{2}(n^2 + n)$. If this expression is also square, we have the Diophantine equation $\frac{1}{2}(n^2 + n) = m^2$. Learning the technique of solving this equation is an excellent introduction to Diophantine analysis. The initial step is to manipulate the equation to get a simpler equation that will be a key to the solution. One way to do it is:

1. Express the equation as $n^2 + n = 2m^2$.
2. Multiply each side by 4: $4n^2 + 4n = 8m^2$.
3. Add 1 to each side: $4n^2 + 4n + 1 = 8m^2 + 1$.
4. Factor $(2n + 1)(2n + 1) = 2(4m^2) + 1$.
5. Let $y = (2n + 1)$ and $x = 2m$.
6. Substituting these terms in the preceding equation produces $y^2 = 2x^2 + 1$.

This is the simplest form of what is called the Pell equation, about which more below. If we can find an integral solution for it, we can easily work backward to find integral values for n and m in the original equation. A standard algorithm for cracking a Pell equation is to express the square root of the coefficient of x (in this case 2) as a continued fraction, then explore its convergents for values of x and y that satisfy the Pell. The technique is too involved to explain here, but interested readers will find a good introduction to it in Chapter 22 of Albert H. Beiler's *Recreations in the Theory of Numbers* or in any good textbook on Diophantine analysis.

It turns out that whenever the coefficient is not a square, the Pell has an infinity of solutions. Since 2 is not a square, there is an infinity of square triangles. The sequence begins 1, 36, 1225, 41616, 1413721, ... The recursive procedure for continuing this sequence is to multiply the last square triangle by 34, subtract the preceding square triangle, then add 2. A nonrecursive formula for the nth square triangle is

$$\frac{[(17 + 12 \sqrt{2})^n + (17 - 12 \sqrt{2})^n - 2]}{32}$$

The irrational numbers in the formula might lead one to suppose that rounding up or down is necessary, but this is not the case. The formula is exact. Substitute any positive integer for n, and the irrationals drop out to give an integral value for the expression. It is surprising how often the problem of finding this formula turns up in the problem departments of mathematical journals, even though it has been shown that the formula goes back to Euler.

Square triangles have many unusual properties. One of the most surprising is that when a simple algorithm is applied, each square triangle gives the sides of an integral right triangle with one leg exactly one unit longer than the other. Let the sides of the right triangle be x and $x + 1$ and z the hypotenuse. Let v be the square root of a square triangle and u its side when represented as a triangle. The procedure is merely to solve these two simultaneous equations:

$$u = z - x - 1$$
$$v = \tfrac{1}{2}(2x + 1 - z)$$

For example, if we take the second square triangle, 36, then $v = 6$, $u = 8$. The above equations give x a value of 20 and z a value of 29. The Pythagorean triplet therefore is 20, 21, 29. Had we used the first square triangle, 1, the algorithm would have provided the familiar 3, 4, 5 right triangle. The third square triangle gives the triplet 119, 120, 169. In this way, all Pythagorean triangles with consecutive legs can be obtained from the square triangles, and of course we can go the other way and derive all the square triangles from consecutive-legged Pythagorean triangles. The simple procedure, or one equivalent to it, explains how Beiler was able to construct his table (pp. 328

and 329 in his book) of the first 100 Pythagorean triangles with consecutive legs. The 100th such triangle has legs that are each expressed by a seventy-seven-digit number. No Pythagorean triangle can have equal legs, but this monstrosity is so nearly isosceles that, as Beiler graphically points out, if its smaller leg is a light-year long, the other leg would be longer by an amount so infinitesimal that the difference between the two legs would be millions of times less than the diameter of a proton.

We turn now to two planar figurate numbers that are not polygonal in the classic sense. The first has received scant attention in the past. The second, so far as I know, has not previously been recognized as a figurate number.

If we arrange points as shown in Figure 3, we have what are known as centered hexagonal numbers to contrast them with the traditional vertex-generated numbers. Let us call them "hexes" for short. As the illustration makes clear at a glance, the formula for the nth hex is $3n(n-1)+1$. It is the sum of three rhombuses, each of sides n and $(n-1)$, plus the single spot in the center. Figure 4 shows that a hex is also the sum of six triangles plus the central spot. The sequence begins 1, 7, 19, 37, 61, 91, 127, 169,. . . . The recursive procedure is to multiply by 2, subtract the preceding number, and add 6.

Suppose we build a hex pyramid of coins, starting with a hex that has 100 coins on the side. On top of this, we put a hex of 99 coins on the side, then one of 98, and so on, until finally we cap the pyramid with a single coin on the

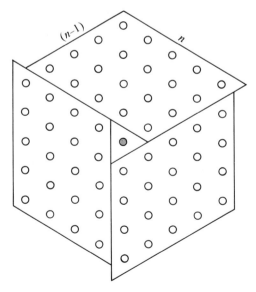

Figure 3 Formula for nth hex number is $[3 \times n(n-1)] + 1$

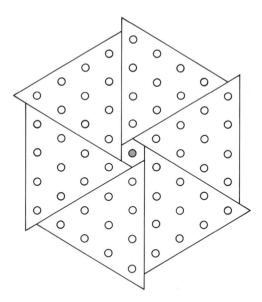

Figure 4   Six triangles plus the center make a hex

central stack. The pyramid is 100 layers high. How many coins are in it? To answer this, we need to know the formula for the sum of the first n hexes. The answer is unexpectedly simple. It is n^3. There are therefore $100^3 = 1,000,000$ coins in the hex pyramid.

It follows from this formula that every hex is the difference between two consecutive cubes. We can demonstrate this elegantly by building a cube, say a $5 \times 5 \times 5$, out of unit cubes. Remove one cube (the first hex) from a top corner. It leaves a $1 \times 1 \times 1$ hole. Around that hole are seven cubes (the second hex). Removing these seven leaves a $2 \times 2 \times 2$ cubical hole. Surrounding this hole are nineteen cubes (the third hex). Removing the nineteen cubes leaves a $3 \times 3 \times 3$ cubical hole, and so on (*see* Figure 5).

Apart from a hex of 1, the first triangular hex is 91, and the first square hex is 169. Readers who know the Pellian technique may enjoy searching for recursive procedures that will generate each of these infinite sequences of numbers and for their nonrecursive formulas. The Pellian for square hexes is $3x^2 + 1 = y^2$, which is solved by finding the convergents of the continued fraction for the square root of 3. The next square hex after 169 is 32761, and the next is 6355441. Are there hexes that are both square and triangular? Is there a cubical hex?

Closely related to hexes are numbers that I have not seen recognized before as figurate, although one often sees them as patterns for drainage holes and for

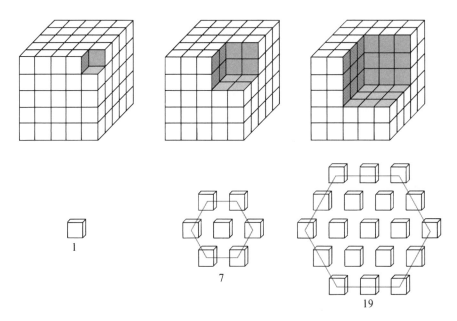

Figure 5 Hexagonal numbers as the difference between consecutive cubes

holes on salt and pepper shakers. Game enthusiasts know them as the patterns of boards for playing Chinese checkers. Let us call them "star" numbers. Figure 6 shows the first four stars, and Figure 7 is a "look-see" proof of the formula for the nth star. Clearly, a star consists of six rhombuses, each n by $(n - 1)$ plus the central spot, or $6n(n - 1) + 1$. A star is also the sum of twelve triangles plus the central spot, as is shown in Figure 8.

The star sequence begins 1, 13, 37, 73, 121, 181, 253, 337, 433, 541,. . . . Adding $12n$ to the nth star produces the next star. A hex contains six triangles. Adding six more triangles to its six sides produces a star; consequently any hex number becomes a star number if we double it and subtract 1. The first n stars add up to $2n^3 - n$. Is this sum ever a square? Yes, but only when $n = 1$ or 169. This was established in 1973 by John Harris, on the basis of results reported by Louis J. Mordell on page 271 of his *Diophantine Equations*.

The first triangular star after 1 is 253. The recursive procedure is to multiply a triangular star by 194, add 60, and subtract the preceding triangular star. The infinite sequence begins 1, 253, 49141, 9533161, 1849384153,. . . . The nonrecursive formula for the nth triangular star is

$$\frac{3[(7 + 4\sqrt{3})^{2n-1} + (7 - 4\sqrt{3})^{2n-1}] - 10}{32}$$

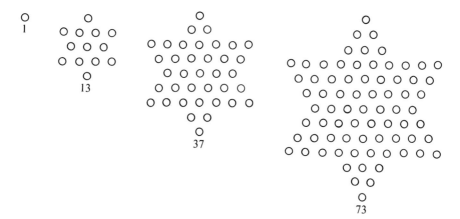

Figure 6 The first four stars

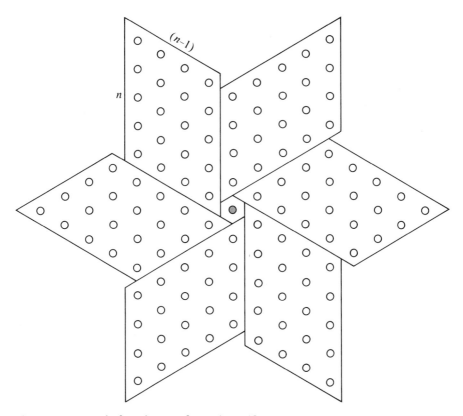

Figure 7 Formula for *n*th star is $[6 \times n(n - 1)] + 1$

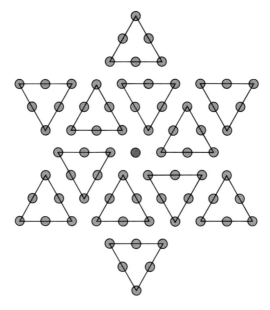

Figure 8 Twelve triangles plus the center make a star

The first square star after 1 is 121. It is the number of holes on the standard Chinese checkers board. The recursive procedure is to multiply a square star by 98, subtract the preceding square star, and add 24. The infinite sequence begins 1, 121, 11881, 1164241, 114083761,. . . . For readers who care to work through the solution of the equation for square stars, $6n(n-1) + 1 = m^2$, I shall say only that this reduces to a solution of $2x^2 + 1 = 3y^2$, where x is the square root of a square star. It can be solved by finding alternate convergents of the square root of $\frac{3}{2}$. The extract nonrecursive formula for the nth square star is

$$\left[\frac{(5 + 2\sqrt{6})^n(\sqrt{6} - 2) - (5 - 2\sqrt{6})^n(\sqrt{6} + 2)}{4}\right]^2$$

One of the most remarkable properties of square stars is that they provide a simple algorithm for producing every number that can be expressed as the sum of two consecutive squares and also as the sum of three consecutive squares. The smallest such number is 365 (the number of days in the year), which equals $13^2 + 14^2$ and also equals $10^2 + 11^2 + 12^2$. The procedure is simply to take any square star greater than 1, triple it, and add 2. The smallest square star greater than 1 is 121. Three times 121, plus 2, is 365. The next square star,

11881, leads to the number 35645, which equals $133^2 + 134^2$ and also equals $108^2 + 109^2 + 110^2$. The third case is $3(1164241) + 2 = 3492725 = 1321^2 + 1322^2 = 1078^2 + 1079^2 + 1080^2$. In each case, the middle term of the triplet of consecutive squares is the original square star.

It is a pleasant exercise, demanding no special skills in number theory, to show that the algorithm always works. Can the reader find a simple proof?

I have been told by Victor Meally that Matila Ghyka, a Romanian mathematician, published some studies of hexes, but I do not know the references. I was unable to find any discussion of stars as such, although their formula turns up in connection with many Diophantine problems. One can raise all kinds of questions about stars that may be easy or difficult to answer. I do not know, for example, if there are stars that are both square and triangular. Since a star's digital root is 1 or 4, and a triangle's digital root must be 1, 3, 6, or 9, we can say that a square triangular star must have a digital root of 1, but that is not of much help.

The general Pell equation, a key to so much of this kind of number analysis, is $ax^2 + 1 = y^2$, where a is a positive integer. It has an infinity of positive integral solutions for x and y, unless a is a square, in which case there are no solutions. As we have seen, when $a = 2$, we have the key to finding square triangles, and when $a = 3$, the key to finding square hexes. The equation was mistakenly named for John Pell, a seventeenth century number theorist, because of a false impression on Euler's part. Pell had nothing to do with the equation. It was known to early Greeks and Hindus, but Pierre Fermat was the first to propose advanced work on it, and the general solution was obtained by John Wallis and others. The classic reference is *The Pell Equation* by E. E. Whitford. The tables in this book can save vast amounts of tedious calculations with continued fractions. As J. A. Lindon has put it elegantly in one of his unpublished mathematical Clerihews:

> To equations simultaneously Pellian
> My approach is Machiavellian.
> Anything goes, rather than resort to such actions
> As covering the walls with continued fractions.

I wish to thank Meally for providing the nonrecursive formulas for square stars and triangular stars, as well as other things in this chapter, and also to thank John Harris and John McKay for additional assistance. I also found *A Handbook of Integer Sequences* by N. J. A. Sloane to be an invaluable tool. I shall say no more about this marvelous reference, except that every recreational mathematician should buy a copy forthwith.

ANSWERS

The problem was to prove that if a star number is multiplied by 3, and 2 is added, the result is a number that can be expressed as the sum of two consecutive squares and also as the sum of three consecutive squares. As explained, square stars are numbers of the form $6n(n-1) + 1 = m^2$, where n and m are positive integers. If the left side, which defines a star, is multiplied by 3, and 2 is added, the result is $18n(n-1) + 3 + 2 = 18n^2 - 18n + 5$. The expression is equal to the sum of two consecutive squares: $(3n-1)^2 + (3n-2)^2$. If the right side of the equation, which defines a square, is multiplied by 3, and 2 is added, the result is $3m^2 + 2$. This equals the sum of three consecutive squares $(m-1)^2 + m^2 + (m+1)^2$.

ADDENDUM

I asked if a hex number could be both triangular and square. The answer is: only hex number 1. The proof was given by Charles M. Grinstead, a mathematician at Swarthmore College, in his paper "On a Method of Solving a Class of Diophantine Equations" (*Mathematics of Computation* 32, July 1978, pp. 936–940).

I also asked if a hex could be a cube. David Chess, Sin Hitotumatu, and Wesley Johnson were the first to tell me how easily it can be shown that the answer is no. Every hex, as I explained, is the difference between two consecutive cubes. The question, therefore, is the same as asking if the formula $(x+1)^3 - x^3 = y^3$ has an integral solution. When this is written $x^3 + y^3 = (x+1)^3$, we see at once that it is a case of Fermat's last theorem when the exponents are 3. This was long ago proved to have no integral solution.

Harvey J. Hindin, in the *Journal of Recreational Mathematics*, showed that the problem of determining all numbers that are both hexes and stars is equivalent to the task of determining when one triangular number is twice another, and also the same as finding Pythagorean triples x, y, z, such that $y = x + 1$. A table in his article provides the first ten star-hexes, an infinite sequence that begins 1, 37, 1261, 42841,. . . . Hindin has calculated the first 15,000 stars and hexes. The values are available from him at 5 Kinsella Street, Dix Hills, NY 11746.

John Harris pointed out in a letter that every star has a digital root (its value modulo 9) of 1 or 4, and that the final pair of digits must be 01, 21, 41, 61, 81, 13, 33, 53, 73, 93, or 37. This rules out any star number containing each of the digits 1 through 9 just once (such a number would have a digital root of 9), and also any "rep-digit" star consisting entirely of repetitions of the same digit.

Harris also showed that every pair of consecutive stars are relatively prime, and every prime divisor of a star is one more or one less than a multiple of 12.

Several readers wrote to say that early unofficial flags of the United States had thirteen stars (for the original thirteen states) arranged in star formation. Note also that this pattern is on the green side of a dollar bill, just above the eagle.

B   I   B   L   I   O   G   R   A   P   H   Y

The Pell Equation. E. E. Whitford. Columbia University Press, 1912.

Recreations in the Theory of Numbers. Albert H. Beiler. Dover, 1964.

Diophantine Equations. Louis J. Mordell. Academic Press, 1969.

A Handbook of Integer Sequences. N. J. A. Sloane. Academic Press, 1973.

"Diophantine Analysis and Fermat's Last Theorem." Martin Gardner, in *Wheels, Life, and Other Mathematical Amusements.* W. H. Freeman and Company, 1983.

"Stars, Hexes, Triangular Numbers, and Pythagorean Triples." Harvey J. Hindin, in *Journal of Recreational Mathematics* 16, 1983–1984, pp. 191–193.

"Figurate Numbers: The Other Alternative." Allan Whitcombe, in *Mathematics in School* (a publication of England's Mathematical Association), September 1986, pp. 40–42. The set of central figurate numbers is the "other alternative" to the vertex figurate numbers.

THREE

. . .

Tangrams, Part 1

"The seven Books of Tan . . .
illustrate the creation of the world
and the origin of species upon a plan
which out-Darwins Darwin, the
progress of the human race being
traced through seven stages . . . up
to a mysterious spiritual state which
is too lunatic for serious
consideration."

—SAM LOYD, *The Eighth Book of Tan*

One of the oldest branches of recreational mathematics has to do with dissection puzzles. Plane or solid figures are cut into various pieces, and the problem is to fit the pieces together to make the original figure or some other figure. The outstanding recreation of this type since the Renaissance is the Chinese puzzle game known as tangrams.

Although tangrams and jigsaw puzzles have a superficial resemblance, they are poles apart in the kinds of challenge they offer. As Ronald C. Read points out in his book *Tangrams: 330 Puzzles,* a typical jigsaw puzzle consists of hundreds of irregularly shaped pieces that fit together in just one way to make a large pattern. Little skill is required, just time and patience. Tangrams have only seven pieces, called tans. They are of the simplest possible shapes and are used to make an infinite variety of tangrams. In creating these figures a heavy demand is made on one's geometrical intuition and artistic ability.

The tans are obtained by slicing a square to produce two large triangles, a middle-size triangle, two small triangles, a square, and a rhomboid (*see* Figure 9). Note that all the corners are multiples of 45 degrees. If a side of the square tan is taken as unity, a side of any tan has one of four lengths: 1, 2, $\sqrt{2}$, and $2\sqrt{2}$.

"At first we are amazed at the unfitness of the shapes . . . with which we are expected to accomplish so much," wrote Loyd, the American puzzle expert. "The number 7 is an obstinate prime which cannot be divided into symmetrical halves, and the geometrical forms . . . with harsh angles, pre-

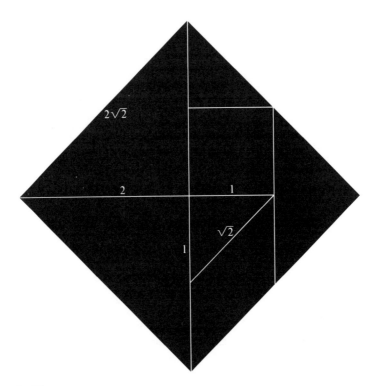

Figure 9   The seven tans

clude the possibility of variety, curves or graceful lines." After working for a while with the tans, however, one begins to appreciate the subtle elegance of the dissection and the richness of its combinatorial possibilities. All kinds of variant dissections, in imitation of tangrams, have been marketed from time to time, but not one has come even close to tangrams in popularity. As with origami, it is the very simplicity of the material and its apparent unfitness for artistic use that lie at the heart of its charm.

Tangram play falls roughly into three major categories:

1. Searching for one or more ways to form a given tangram, or for an elegant proof of the impossibility of forming a tangram.
2. Finding ways to depict, with maximum artistry or humor, or both, silhouettes of animals, human figures, and other recognizable objects.
3. Solving a variety of problems in combinatorial geometry that are posed by the seven tans.

Many books, and even a few encyclopedias, declare that tangram play is about 4000 years old. In my *Scientific American* column for September 1959, I called tangrams the oldest of dissection games and said that the Chinese had been amusing themselves with it for several thousand years. This is totally wrong. The man responsible for the myth is none other than Sam Loyd. In 1903, when Loyd was sixty-one and at the height of his fame, he published a little book (now extremely rare) called *The Eighth Book of Tan*. No Western book on tangrams has been more original or influential. In addition to containing hundreds of excellent new figures, Loyd invented a preposterous legend about the pastime's origin. It was the greatest hoax in the history of puzzledom, and the number of intelligent people taken in by it rivals the number of scholars who accepted H. L. Mencken's spurious history of the bathtub.

"According to the late Professor Challenor," Loyd wrote, "whose posthumous papers have come into the possession of the writer, seven books of Tangrams, containing one thousand designs each, are known to have been compiled in China over 4000 years ago. These books are so rare that Professor Challenor says that during a forty years' residence in China he only succeeded in seeing perfect editions of the first and seventh volumes, with stray fragments of the second.

"In this connection it may be mentioned that portions of one of the books, printed in gold leaf upon parchment, were found in Pekin by an English soldier who sold it for £300 to a collector of Chinese antiquities, who kindly furnished some of the choicest designs presented in this work."

According to Loyd, Tan was a legendary Chinese writer who was wor-shiped as a deity. He arranged the patterns in his seven books to display seven stages in the evolution of the earth. His tangrams begin with symbolic repre-sentations of chaos and the yin-yang principle. These are followed by primi-tive forms of life, then the figures proceed up the evolutionary tree through fish, birds, and animals to the human race. Scattered along the way are tan-grams of human artifacts such as tools, furniture, clothing, and architecture. Loyd inserts remarks by Confucius, a philosopher called Choofootze, a com-mentator named Li Hung Chang, and his mythical Professor Challenor. Chang is quoted as saying that he knew all the figures in the seven books of Tan before he could talk. And there are references to a "well-known" Chinese saying about "the fool who would write the eighth book of Tan."

All of this, of course, was sheer fabrication. When Henry Ernest Dudeney, Loyd's British counterpart, wrote an article on tangrams for *The Strand Maga-zine* (November 1908), he soberly repeated Loyd's legendary history. This aroused the curiosity of Sir James Murray, the distinguished lexicographer and an editor of the *Oxford English Dictionary,* who made inquiries through one of his sons, then teaching at a Chinese university. Oriental scholars had never heard of Tan or even the word tangrams. The game, Murray informed Dudeney, is known in China as *ch'i ch'iao t'u,* meaning "seven-ingenious plan" or, less literally, "clever puzzle of seven pieces."

Murray could find no record of the word tangram earlier than in an 1864 Webster's dictionary. It had been coined about 1850, Murray guessed, by an American who probably combined *tang,* a Cantonese word for "Chinese," with the familiar suffix *-gram,* as in anagram or cryptogram. A different theory about the name has recently been advanced by Peter Van Note in his introduction to a Dover reprint of Loyd's fanciful book. Chinese families who live on riverboats are called *tanka,* and *tan* is a Chinese word for prostitute. American sailors, taught the puzzle by tanka girls, may have called it tangrams—the puzzle of the prostitutes.

When Dudeney reported Murray's opinions, in *Amusements in Mathematics* (pp. 43–46), he may have deliberately added a hoax of his own. An American correspondent, Dudeney writes, had told him that he owned a Chinese set of mother-of-pearl tans with an accompanying rice-paper booklet of more than 300 figures. The correspondent was puzzled by a mysterious inscription on the front page that he said he had tried to have translated, but no Chinese to whom he had shown it had been willing or able to read it. Dudeney repro-duced the inscription and asked the reader for help. We do not know what the response was to this request, but Read, who owns a copy of the same booklet, had no difficulty clearing up the mystery. The inscription is nothing more than a caption under the tangrams of two men. The caption reads: "Two men facing each other and drinking. This shows the versatility of the seven-piece puzzle."

No one knows when tangrams originated. The earliest reference known is a book published in China in 1803. Its title, *The Collected Volume of Patterns of the Seven-Piece Puzzle,* suggests earlier books. Most scholars believe the game originated in China about 1800, became an Oriental craze, and then spread rapidly to the West. The earliest Western books, says Read, were little more than copies of Chinese rice-paper booklets. The Western books even copied errors in the Chinese illustrations.

One of the earliest English books on tangrams, originally owned by Charles Lutwidge Dodgson (better known as Lewis Carroll), came into Dudeney's possession. It is called *The Fashionable Chinese Puzzle* and was first published in New York in 1817. Dudeney quotes from it a passage stating that the game was a favorite of "ex-Emperor Napoleon, who, being now in a debilitated state, and living very retired, passes many hours a day in thus exercising his patience and ingenuity." This too is an unsupported statement, undoubtedly false. The puzzle is said by Loyd to have been a favorite of John Quincy Adams's and Gustave Doré's, although I know of no basis for either assertion. We do know that Edgar Allan Poe enjoyed the game, because his imported set of carved ivory tans is owned by the New York Public Library. An anonymous French work, *Recueil des Plus Jolie Jeux de Société* (1818), may be a translation of Dodgson's English book, or vice versa. I have not seen a copy of either. An 1817 American book bears the title *Chinese Philosophical and Mathematical Trangram.* "Trangam" was an old English word for a trinket, toy, or puzzle. Samuel Johnson misspelled it as "trangram," and the spelling persisted in later dictionaries. Did the book's anonymous author revive an obsolete word that later evolved to "tangram," or did he misspell "tangram," a word already in use? One mystery novel, *The Chinese Nail Murders* by the Dutch diplomat and Orientalist Robert Van Gulik, is woven around a set of tangram patterns.

Poe's tans are shown in Figure 10. The delicate filigree carving is characteristic of the old Chinese ivory tans. Note that the pieces pack into a square box in two layers. The two layers are squares of equal size, so that putting away the tans is a puzzle in itself. In nineteenth-century China, where tangrams were popular among adults (it is considered a child's pastime in the Far East today), the pieces were made in many sizes and from many different materials. Dishes, lacquer boxes, and even small tables were given the shapes of the tans.

So much for the historical background. Let us turn now to the first of the three categories of tangram play: solving given figures. Figure 11 shows a dozen interesting shapes on which the reader is invited to try his skill. Each

Figure 10   Edgar Allan Poe's carved ivory tangram set

requires all seven pieces. The rhomboid, the only asymmetrical tan, may be placed either side up. One figure in the illustration is not possible. Can the reader identify it and prove its impossibility?

The paired tangrams in Figure 12 are samples of delightful paradoxes introduced by Loyd. (The first three pairs were devised by Loyd, the fourth pair was devised by Dudeney.) Although the figure at the right in each case seems to be exactly the same as its mate, except for a missing portion, each is made with all seven tans!

The tangrams in Figure 13 are not intended as patterns to be solved, but as illustrations of the second category of play: creating artistic and amusing pictures. (I confess responsibility for the Nixon caricature.) "One remarkable thing about . . . Tangram pictures," wrote Dudeney, "is that they suggest

Figure 11 Which tangram is impossible?

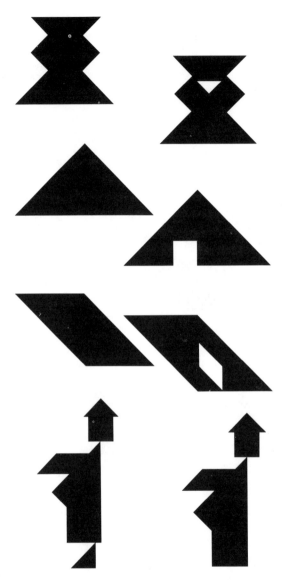

Figure 12 Tangram paradoxes

to the imagination such a lot that is not really there. Who, for example, can look . . . at Lady Belinda . . . without soon feeling the haughty expression . . . ? Then look again at the stork, and see how it is suggested to the mind that the leg is actually much more slender than any one of the pieces employed. It is really an optical illusion. Again, notice in the case of the yacht how, by leaving that little angular point at the top, a complete mast is suggested. If you place your tangrams together on white paper so that they do not quite touch one another, in some cases the effect is improved by the white lines; in other cases it is almost destroyed."

One can mix two or more sets of tans to produce more elaborate figures. Dudeney gives a number of these "double tangrams" in *536 Puzzles and Curious Problems* (pp. 221–222), and others will be found in Read's book. I agree with Read, however, when he writes: "With fourteen pieces to play around with, one cannot help but feel that it should be possible to arrive at a reasonable likeness of just about anything. Consequently, the sense of achievement that one gets on producing a recognizable cow, sailing boat, human figure, or what have you, from a mere seven pieces, is quite lacking."

Combining two related tangrams, each made with seven tans, is a different matter. Four classic examples, all devised by Loyd, are a woman pushing a baby carriage, a runner being tagged out at home plate, two Indian braves, and a man with a wheelbarrow (*see* Figure 14). Note that the man and the wheelbarrow are identical tangrams except for orientation.

The third category of tangram play, solving combinatorial problems, is the most interesting of all to mathematicians. There have been some remarkable

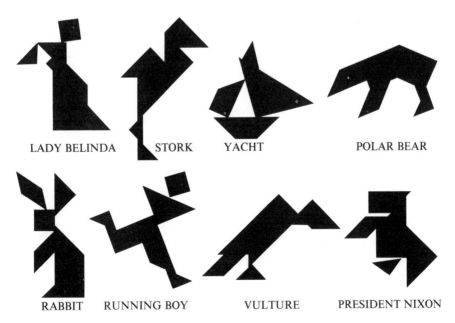

LADY BELINDA   STORK   YACHT   POLAR BEAR

RABBIT   RUNNING BOY   VULTURE   PRESIDENT NIXON

Figure 13 Tangram pictures

Figure 14 Double tangrams by Sam Loyd

contributions made to this field by Read, a specialist in graph theory at the University of Waterloo, and by E. S. Deutsch, a computer scientist with P. S. Ross and Partners in Toronto. Some of their results will be presented in the next chapter. To whet the reader's appetite, here are two problems that will be answered in the next chapter.

1. How many different convex polygons can be formed with the seven tans? There must not be any "windows" in the figures. Rotations and reflections are not, as is customary, considered to be different. Because all three-sided polygons are convex, and no nonconvex polygon of four sides can be made with all seven tans, answering this question also gives the number of three- and four-sided polygons. It is easy to see that only one triangle is possible (since corners must be multiples of 45 degrees, the triangle must be a right isosceles one), but finding all the higher convex polygons is a bit tricky.
2. How many different five-sided polygons can be made?

ANSWERS

The impossible tangram in Figure 11 is the square with the central square hole. The two large triangles can go only in opposite corners. The square tan must go in a third corner and the rhomboid must touch the fourth corner, but now there is no spot for the middle-size triangle.

B I B L I O G R A P H Y

The Bibliography following Chapter 4 (p. 54) is also for this chapter.

FOUR

·
·
·

Tangrams, Part 2

"The formation of designs by means
of seven pieces of wood . . . known
as tans . . . is one of the oldest
amusements in the East. Many
hundreds of figures representing
men, women, birds, beasts, fish,
houses, boats, domestic objects,
designs, etc., can be made, but the
recreation is not mathematical, and I
reluctantly content myself with a
bare mention of it."

—W. W. ROUSE BALL, *Mathematical
Recreations and Essays*

Not mathematical? Ball wrote without giving the matter much thought. In
this chapter we consider some nontrivial combinatorial problems presented by
the seven tans.

A question that arises at once is: How many sides can a tangram formed with
all seven tans have? Although the answer is obvious, it seems to have been first

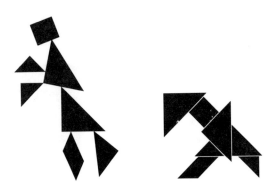

Figure 15 Improper tangram (left) and proper tangram (right), each with twenty-three sides

stated by Harry Lindgren in his 1968 article "Tangrams." This is how Ronald C. Read, a mathematician at the University of Waterloo, puts it in a lengthy communication to me from which I shall be quoting liberally:

"Tangrams have 23 sides between them, so that a tangram like that (at the left in Figure 15) will clearly have this number of sides. On the other hand, tangrams with pieces joined only at a corner are mathematically uninteresting . . . . " Let us make a rule, Read proposes, that a tangram must have a perimeter topologically equivalent to a circle, that is, it must not self-intersect. Read calls these "proper tangrams." How many sides can a proper tangram have? Again the answer is twenty-three. The proof is supplied by the figure of a bowing man shown at the right in Figure 15 and by almost endless other examples.

The proper tangrams contain an important subset that Read calls "snug tangrams." In order to understand the meaning of snug, draw lines on all tans (except the two small triangles) to create sixteen identical right-isosceles triangles with unit legs (*see* Figure 16). A snug tangram is a proper tangram formed so that, where two tans are in contact, the sides of the small right triangles match exactly, either leg to leg or hypotenuse to hypotenuse. All convex tangrams are snug, and so are many of the traditional figures (*see* Figure 17).

Snugness, by the way, is characteristic of technology in the Orient, where the dimensions of houses, furniture, and so on tend to be exact multiples of a basic length. The Japanese building industry, I am told, is one of the most efficient in the world because Japanese lumber is standardized in lengths that are multiples of a basic "mat" length.

In addition to snugness of fit, Read adds two more limitations: A snug tangram must be simply connected (all in one piece), and there must be no

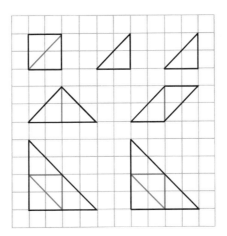

Figure 16   The seven tans

interior holes, including holes that touch the perimeter at one or more single points. It is convenient to diagram snug tangrams on graph paper so that all integral edges are on the orthogonals of the matrix. All diagonal edges will then be multiples of $\sqrt{2}$ and therefore irrational. This suggested to Read a very pretty problem: How many snug tangrams have *all* sides irrational? Such tangrams, if diagrammed on graph paper, would have every side running diagonally.

The tans between them have a total of thirty side segments, Read continues, but "whenever we place two pieces together, the two sides that abut are lost to the perimeter, and it can happen that we lose more than two. Furthermore, in order that the resulting tangram shall be connected, there must be at least six lines along which two pieces come together. Hence we cannot avoid losing 12 segments, therefore the total number of segments on the outside cannot be

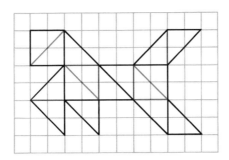

Figure 17   A snug tangram of a dog with eighteen sides

more than 18. Each side of a snug tangram consists of at least one segment, hence the total number of sides cannot exceed 18 either." The dog in Figure 17 proves that a snug tangram can have this maximum total of sides.

The number of proper tangrams obviously is infinite. You have only to observe that two tans can abut in an infinite number of positions. If the question is confined to certain kinds of tangram, however, interesting problems in combinatorial enumeration arise. For example, how many convex tangrams are there? A convex tangram is a polygon in which all corner angles are less than 180 degrees. That there are only thirteen was proved in 1942 by Fu Tsiang Wang and Chuan-Chih Hsiung in "A Theorem on the Tangram" in *The American Mathematical Monthly*. The thirteen are shown in Figure 18. If mirror images count as different convex tangrams, then there are eighteen. The eighteen convex tangrams appear in Chinese tangram books with solutions showing that all can be made without turning over the asymmetric rhomboid tan. (Interior lines are omitted in the illustration, in case some readers might enjoy solving them.)

The thirteen convex tangrams include all three- and four-sided polygons that can be made with the seven pieces. As stated in the previous chapter, nonconvex quadrilaterals are not possible. (Can you prove this? Hint: The four interior angles of such a four-sided figure would have to be three angles of 45 degrees and one of 225 degrees, and the figure would have to consist of sixteen isosceles right triangles congruent with the small triangular tan.) Five-sided polygons, made with the tans, *can* be nonconvex. How many pentagons, both convex and nonconvex, Lindgren asked, can be made with the seven tans?

At this spot in my 1974 column, I gave a proof that there are sixteen snug pentagons and two "loose" pentagons, making eighteen in all. Unfortunately, there was a flaw in the proof, which readers quickly discovered. For months I was flooded with hundreds of letters from readers of all ages who had found pentagons not among the eighteen that I have given in an illustration.

I kept tab of the number of different pentagons sent by readers until it reached a maximum of twenty-two snug and thirty-one loose ones, or fifty-three in all. Only two readers, each working independently and by hand, found all fifty-three. They are Allan L. Sluizer of Northbrook, Illinois, and Åke Lindgren of Uppsala, Sweden. In 1975 Dr. I. Takeuchi at the Institute for Electrical Engineers, Musashimo, near Tokyo, confirmed the fifty-three figures by a computer program. In 1976, unaware of Takeuchi's program, Michael Beeler of Cambridge, Massachusetts, wrote a program that gave the same results. Beeler's drawings of the fifty-three pentagons are shown in Figure 19.

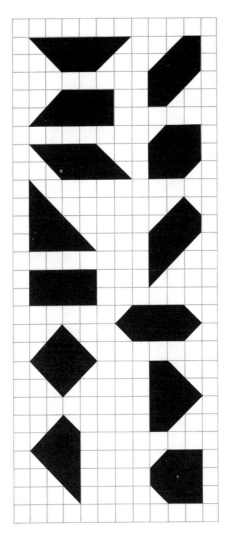

Figure 18 The thirteen convex tangrams

One might now wonder how many tangram polygons have six sides, or seven or more, but (as Read points out) the question is easily answered. For $n = 6$ through 23 there is an infinity of n-sided polygons. You have only to glance at pentagon 28 in Figure 19 to see that by sliding the large triangle on the left along the hypotenuse of the other large triangle, you can create an infinity of hexagons. How many snug hexagons are there? Although it is a finite number, as far as I know the number has not yet been determined.

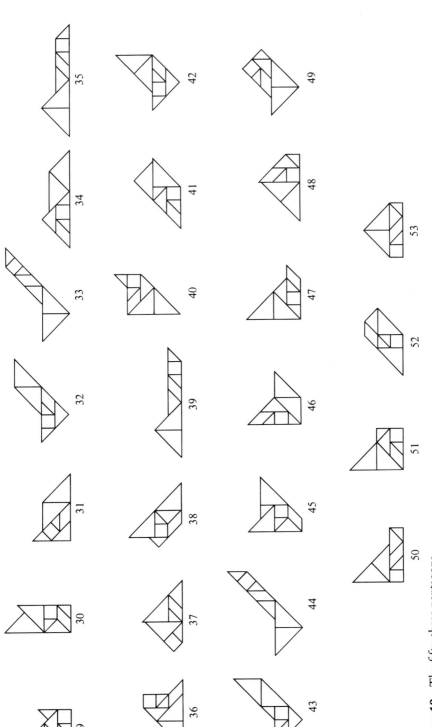

Figure 19 The fifty-three pentagons

There are, of course, only a finite number of snug tangrams, but the exact number (Read calls it the snug number) also is far from known. Read has devised an ingenious procedure by which a computer could be programmed to count the number, but he estimates that it is well into the millions, and no such program has yet been written. Unfortunately, the details of Read's procedure are too complex to give here. A simpler problem, using exactly the same procedure, has been solved, however. Read defines a minitangram as one made with the five pieces that remain after leaving out the two large triangles. The problem of finding all snug minitangrams is so much simpler than finding all snug tangrams that Read was able to write a computer program to solve it and ran it on, appropriately enough, a minicomputer at the University of Waterloo. It took only half an hour, and the count was 951. The computer was hooked up to a display terminal, so that it drew pictures of all the minitangrams.

Read's programs are designed only for enumeration, not for solving individual tangrams. Is it possible to write a program that will inspect any given tangram and search for at least one solution? Yes, such a program has been developed and published by E. S. Deutsch, a computer scientist. In theory it is possible to write a program that will systematically examine all possible ways the tans fit a given tangram and then print all the solutions, but the complexity of such a program is so great that no one has yet attempted it. Deutsch's program is not of this type. It is heuristic, which means that it goes about solving a tangram in much the same way a person does: by applying a series of tentative tests, examining the feedback, backtracking and trying again when no solution is found, and continuing until it either discovers a solution or gives up. The program seldom fails, taking an average time of about two seconds to solve a tangram.

The program starts by examining the tangram's perimeter, noting the edge lengths, and the angles at each corner. It then attempts to separate the tangram into two or more subtangrams. For instance, if two portions of the tangram meet at a point, each portion clearly is a separate tangram. If, say, a rabbit has two ears formed by the two small triangles, each ear meeting the head at a point, the program instantly identifies the two pieces, removes them and goes to work on the subtangram that remains. If the tangram does not have portions meeting at points, the program explores ways of dividing it into subtangrams by extending edges from a corner into the figure. In many cases the internal extension of an edge clearly divides the tangram into subtangrams; in other cases the extension is only a possible line of division.

After the program's preliminary exploration, it applies a series of heuristic tests until it finds a way of fitting tans into a subtangram or a possible subtangram. When a fit is found, the subtangram is extracted and the program turns to what remains. The tests are ranked in order of their efficacy, so that the strongest tests can be applied first, then the next strongest, and so on. If no solution is obtained, the program backtracks and begins with the second test. It is impossible to describe the program in more detail, but interested readers will find it fully explained, with flow charts and examples, in "A Heuristic Solution to the Tangram Puzzle," by E. S. Deutsch and Kenneth C. Hayes, Jr., in *Machine Intelligence 7.* A somewhat similar program was developed in 1972 by Ejvind Lynning, a Danish student working with Jacques Cohen, a physicist at Brandeis University.

Snug tangrams are, for obvious reasons, usually more difficult to solve (by a person or a computer) than nonsnug figures, and the difficulty tends to increase as the number of sides decreases. One might suppose a pattern with only one solution would be harder to solve than one with many, but that is not the case. A pattern in which the tans touch only at points has only one solution, but it is an immediately obvious one, and there are patterns with a large number of solutions that are among the most difficult.

The construction of tangrams with "holes" raises many curious new problems. It is not hard to form a square hole of area 4 or a triangular hole of area 2 that does not touch the border, or two triangular holes of areas 1 and $\frac{1}{2}$ that do not touch each other or the border (*see* Figure 20); ("touch" includes touching at a single point). Can the reader find a way to make exactly two holes, each 1 × 1 square, that do not touch each other or the perimeter? Or two holes, under the same provisos, one a triangle and one a square, each with an area of 1? They are not difficult tasks, but here are two more that I set for myself and found much harder: (1) Form just three holes, two triangular and one square, that do not touch one another or the border. (2) Form just three holes, two rectangular and one triangular, that do not touch one another or the border. Apparently it is not possible for three holes of this type to be all rectangular or all triangular, or for two triangular holes each to have an area of 1.

The "farm problem" is another unsolved hole problem. What is the largest hole not touching the border that can be inside a tangram? The solution is a limit that cannot be reached, but one can come as close to it as one wishes. (The best I can obtain is the limit 10.985+.) How many sides can a single hole that is simply connected and not touching the border have? The maximum surely is thirteen. What is the largest "farm" not touching the border that is square? Rectangular? Triangular?

Another unexplored type of tangram problem is finding ways of transforming one tangram into another with the fewest number of moves. A move

Figure 20 Tangrams with holes

consists in altering the position of a set of one or more tans without disturbing the pattern of the set. For example, the large square tangram can be changed to the large triangle or the large rhomboid in one move, or to the 2 × 4 rectangle in three moves. As Read notes in his book, the square can be changed to a 3 × 3 square with a missing 1 × 1 corner in just two moves.

Still another area of tangram play open to exploration is the devising of competitive games that use one or more sets of tans. The only tangram game I have seen in books in English is the party game of giving each guest a set of tans and awarding prizes to those who are the first to make each of a series of displayed patterns. Read's concept of snugness suggests a variety of two-person games. Here are three that occurred to me. In playing them it is a good plan to mark the middle of the long edges to facilitate placing the tans snugly.

1. Snuggle up. Begin with the tans forming the large triangle, the square, or any four-sided polygon. Players alternate moves. A move consists in changing the position of a single tan to form a new snug tangram that has more sides than the preceding one. The first player who is unable to make a move loses.

2. Snuggle down. Same as above, except that the initial tangram is an eighteen-sided snug tangram and each move must decrease the number of sides. As in the preceding game, you cannot move a piece that leaves or forms a hole or that divides the figure into parts that touch only at points. Both games end quickly. Because a snug tangram must have at least three sides and no more than eighteen, a game cannot last beyond fifteen moves. Games that go a full fifteen moves are possible.

3. Snuggle up and down. Start with a ten- or eleven-sided snug tangram. One player must increase the number of sides on each play, the other must decrease the number. The same piece may not be moved twice in succession. Each player keeps a record of his increases and decreases, and the first to score 30 wins. If a player cannot move, he loses. If the up

player makes an eighteen-sided tangram, he wins. If the down player makes a four- or three-sided tangram, he wins. This game lasts considerably longer than the other two, and it often takes unexpected turns. One player can get far ahead in scoring only to discover, just before he expects to make a winning move, that no move is available.

In all three games it is good to keep a running record of the number of sides because it is easy to forget the number, and time is lost by repeated counting. A cribbage board is a convenient device not only for recording the number of sides but also for keeping score in the up-and-down game.

ANSWERS

Solutions to the four hole problems are shown in Figure 21. At the top left in the illustration is a way of making two unit square holes. At the top right are two holes, one a unit square and the other a triangle of area 1. At the left on the bottom of the illustration is a way of making a square hole and two triangular

Figure 21 Solutions to hole problems

holes. (The fit is extremely close. The top triangle's hypotenuse is longer than the length it must span by only .121 + .) At the right on the bottom are two rectangular holes and one triangular hole.

Read has proved that a tangram that is snug except for one or more holes cannot have more than one hole if the holes do not touch one another or the border. The smallest possible hole is equal to a small triangular tan. No matter how two such holes are placed so that they do not touch, at least seventeen triangles are required to isolate them from a tangram border. Since the seven tans are made of sixteen such triangles, completing the required tangram is impossible. For nonsnug tangrams it appears that no more than three holes not touching one another or the border are possible.

The square is the only snug tangram with all its sides irrational. This is how Read proves it. As I explained in the previous chapter, an irrational tangram drawn on graph paper with the unit square tan oriented orthogonally would have all its edges making 45-degree angles with the matrix lines. All corners clearly must be either 90 degrees or 270 degrees. Since each side is a multiple of $\sqrt{2}$ and the total area is 8, it follows that any irrational snug tangram will be a tetromino composed of four squares, each $\sqrt{2}$ on the side.

There are five tetrominoes. One of them, the square, we know can be formed. Each of the other four is easily proved impossible by placing the square tan in each of three possible positions (*see* Figure 22) and then exploring ways of completing the tetromino. The first tetromino is ruled out at once because there is no way to place the two large triangles. In the other three cases, for each position of the square tan, there are at most only four ways to place the two large triangles. In every case, after the square and two large triangles are placed, there is no spot for the rhomboid. Thus the square is the only possible irrational snug tangram.

My solution of the farm problem, with a limit of 10.985 + , is shown in Figure 23.

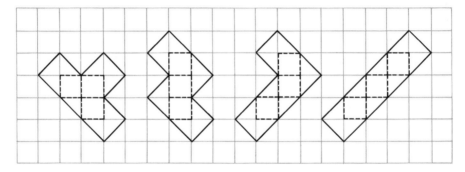

Figure 22 Impossibility proof for the nonsquare tetrominoes

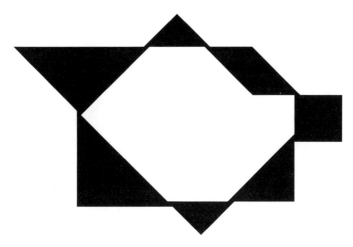

Figure 23   A solution to the farm problem

ADDENDUM

Scores, perhaps hundreds, of puzzles similar to tangrams, but with different dissections of squares, rectangles, circles, and other shapes, have been described in books and articles and manufactured around the world. Some of the marketed variants are pictured in Professor Hoffmann's pioneer book on mechanical puzzles, in *Creative Puzzles of the World* by Pieter Van Delft and Jack Botermans, and in *Puzzles Old and New* by Jerry Slocum and Botermans.

Of special interest, because it predates any known Chinese publication, is a 32-page book published by Kyoto Chobo in Japan in 1742. It gives forty-two patterns to be made with the seven pieces obtained by dissecting a square as shown in Figure 24. The book's title translates as *The Ingenious Pieces of Sei Shonagon*. (Sei Shonagon was a court lady of the late tenth and early eleventh centuries, who wrote the famous *Pillow Book*.) Nothing is known about the author, who uses the pseudonym Ganrei-Ken. It is highly unlikely that Sei Shonagon knew of the puzzle.

Shigeo Takagi, a Tokyo magician, was kind enough to send me a photocopy of this rare book. Unlike the Chinese tans, the Shonagon pieces will form a square in two different ways. Can you find the second pattern? The pieces also will make a square with a central square hole in the same orientation. With the Chinese tans it is not possible to put a square hole anywhere inside a large square.

Richard Reiss, a professor of English at Southeastern Massachusetts University, sent a good proof that no four-sided convex polygon can be made with all

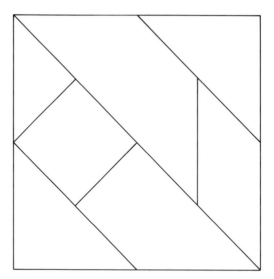

Figure 24 The Sei Shonagon pieces

seven Chinese tans. Peter Van Note proposed the following three tasks based on forming two congruent replicas of a tan:

1. It is possible to make one large square. Use the seven pieces to make two congruent small squares.
2. A large isosceles right triangle is possible. Use the seven pieces to make two congruent isosceles right triangles.
3. A large rhombus is possible. Van Note could not prove it, but he is convinced that the seven pieces cannot make two congruent rhombuses.

John H. Conway wrote from Cambridge University to pose an interesting unsolved problem. What would be the shapes of an "optimal" set of tans — that is, seven convex polygons that will form the largest number of distinct convex polygons?

Figure 25 is taken from Joost Elffers's and Michael Schuyt's marvelous book. The pattern is made with the traditional Chinese tans.

Karl Fulves, author of many books on magic, made the following suggestion for an amusing trick. It involves the tangram paradox at the bottom of

Figure 12. You secretly add to a set of tans a third small triangle. Make the man with the feet, using the pattern shown on the right, and the extra triangle for the feet. Now use sleight of hand to "vanish" one of the small triangles (or just put it in your pocket), then form the man with the foot again, using the pattern on the left. If no one has bothered to count pieces, it seems as if the vanished piece has mysteriously returned. Similar tricks can, of course, be performed with other paradoxical pairs.

Several proposals have been made for using two sets of tangrams to play a board game similar to Solomon W. Golomb's pentomino game. Game inventor Sidney Sackson recommends a 6 × 6 checkerboard, its squares the size of the square tan. Each of two players has a set of seven tans. Players take turns placing any tan on the board, wherever they wish, provided the tan's corners fall on the board's lattice points. The person unable to place a tan loses. Many variations in rules are possible, and larger boards can be used for more than two players.

Figure 25 A tangram to solve

B   I   B   L   I   O   G   R   A   P   H   Y

Puzzles Old and New. "Professor Hoffmann" (pseudonym of Angelo John Lewis). F. Warne, 1893.

Amusements in Mathematics. Henry Ernest Dudeney. Nelson, 1917.

The Tangram Book. F. G. Hartswick. Simon and Schuster, 1925.

"A Theorem on the Tangram." Fu Tsiang Wang and Chuan-Chih Hsiung, in *American Mathematical Monthly* 49, November 1942, pp. 596–599.

The Chinese Nail Murders. Robert Van Gulik. Harper & Row, 1961.

Tangrams: 330 Puzzles. Ronald C. Read. Dover, 1965.

Tangram Teasers. R. C. Bell. Newcastle-upon-Tyne, England (privately published), 1965.

Tangrams. Peter Van Note. Charles E. Tuttle, 1966.

536 Puzzles and Curious Problems. Henry Ernest Dudeney. Scribner's, 1967.

The Eighth Book of Tan. Sam Loyd. Dover, 1968 (reprint of 1903 edition).

"Tangrams." Harry Lindgren, in *Journal of Recreational Mathematics* 1, July 1968, pp. 184–192.

Tangramath. Dale Seymour. Creative Publications, 1971.

A Tangram Tale. William Cameron. Brockhampton Press, 1972.

"A Heuristic Solution to the Tangram Puzzle." E. S. Deutsch and Kenneth C. Hayes, Jr., in *Machine Intelligence* 7. Bernard Meltzer and Donald Michie, eds. Wiley, 1972.

The Fun with Tangrams Kit. Susan Johnson. Dover, 1977.

Creative Puzzles of the World. Pieter Van Delft and Jack Botermans. Abrams, 1978.

Tangrams. Joost Elffers and Michael Schuyt. Abrams, 1979.

Puzzles Old and New. Jerry Slocum and Jack Botermans. Plenary (Holland), 1986, distributed in the United States by the University of Washington Press, Seattle.

FIVE

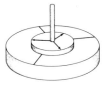

.
.
.
.

Nontransitive Paradoxes

"I have just so much logic, as to be
able to see . . . that for *me* to be too
good for *you,* and for *you* to be too
good for *me,* cannot be true at once,
both ways."

— ELIZABETH BARRETT, *in a letter to*
Robert Browning.

Whenever a relation R that applies to xRy and yRz also applies to xRz, the
relation is said to be transitive. For example, "less than" is transitive among all
real numbers. If 2 is less than π, and the square root of 3 is less than 2, we can be
certain that the square root of 3 is less than π. Equality also is transitive: if $a = b$
and $b = c$, then $a = c$. In everyday life such relations as "earlier than," "heavier
than," "taller than," "inside of," and hundreds of others are transitive.

It is easy to think of relations that are not transitive. If A is the father of B and
B is the father of C, it is never true that A is the father of C. If A loves B and B
loves C, it does not follow that A loves C. Familiar games abound in transitive

rules (if poker hand *A* beats *B* and *B* beats *C,* then *A* beats *C*), but some games have nontransitive (or intransitive) rules. Consider the children's game in which, on the count of three, one either makes a fist to symbolize "rock," extends two fingers for "scissors," or all fingers for "paper." Rock breaks scissors, scissors cut paper, and paper covers rock. In this game the winning relation is nontransitive.

Occasionally in mathematics, particularly in probability theory and decision theory, one comes on a relation that one expects to be transitive but that actually is not. If the nontransitivity is so counterintuitive as to boggle the mind, we have what is called a nontransitive paradox.

The oldest and best-known paradox of this type is a voting paradox sometimes called the Arrow paradox after Kenneth J. Arrow because of its crucial role in Arrow's "impossibility theorem," for which he shared a Nobel prize in economics in 1972. In *Social Choice and Individual Values,* Arrow specified five conditions that almost everyone agrees are essential for any democracy in which social decisions are based on individual preferences expressed by voting. Arrow proved that the five conditions are logically inconsistent. It is not possible to devise a voting system that will not, in certain instances, violate at least one of the five essential conditions. In short, a perfect democratic voting system is in principle impossible.

As Paul A. Samuelson has put it: "The search of the great minds of recorded history for the perfect democracy, it turns out, is the search for a chimera, for a logical self-contradiction . . . . Now scholars all over the world — in mathematics, politics, philosophy, and economics — are trying to salvage what can be salvaged from Arrow's devastating discovery that is to mathematical politics what Kurt Gödel's 1931 impossibility-of-proving-consistency theorem is to mathematical logic."

Let us approach the voting paradox by first considering a fundamental defect of our present system for electing officials. It frequently puts in office a man who is cordially disliked by a majority of voters but who has an enthusiastic minority following. Suppose 40 percent of the voters are enthusiastic supporters of candidate *A.* The opposition is split between 30 percent for *B* and 30 percent for *C. A* is elected even though 60 percent of the voters dislike him.

One popular suggestion for avoiding such consequences of the split vote is to allow voters to rank all candidates in their order of preference. Unfortunately, this too can produce irrational decisions. The matrix in Figure 26 (left) displays the notorious voting paradox in its simplest form. The top row shows that a third of the voters prefer candidates *A, B,* and *C* in the order *ABC.* The middle row shows that another third rank them *BCA,* and the bottom row shows that the remaining third rank them *CAB.* Examine the matrix carefully

VOTERS / RANK ORDER

	1	2	3
1/3	*A*	*B*	*C*
1/3	*B*	*C*	*A*
1/3	*C*	*A*	*B*

TEAMS

	D	*E*	*F*
A	8	1	6
B	3	5	7
C	4	9	2

Figure 26 The voting paradox (left) and the tournament paradox based on the magic square (right)

and you will find that when candidates are ranked in pairs, nontransitivity rears its head. Two-thirds of the voters prefer *A* to *B*, two-thirds prefer *B* to *C*, and two-thirds prefer *C* to *A*. If *A* ran against *B*, *A* would win. If *B* ran against *C*, *B* would win. If *C* ran against *A*, *C* would win. Substitute proposals for men and you see how easily a party in power can rig a decision simply by its choice of which paired proposals to put up first for a vote.

The paradox was recognized by the Marquis de Condorcet and others in the late eighteenth century, and is known in France as the Condorcet effect. Lewis Carroll, who wrote several pamphlets on voting, rediscovered it. Most of the early advocates of proportional representation were totally unaware of this Achilles' heel; indeed, the paradox was not fully recognized by political theorists until the mid-1940s, when Duncan Black, a Welsh economist, rediscovered it in connection with his monumental work on committee decision making. Today the experts are nowhere near agreement on which of Arrow's five conditions should be abandoned in the search for the best voting system. One surprising way out, recommended by many decision theorists, is that when a deadlock arises, a "dictator" is chosen by lot to break it. Something close to this solution actually obtains in certain democracies, England for instance, where a constitutional monarch (selected by chance in the sense that lineage guarantees no special biases) has a carefully limited power to break deadlocks under certain extreme conditions.

The voting paradox can arise in any situation in which a decision must be made between two alternatives from a set of three or more. Suppose that *A*, *B*, and *C* are three men who have simultaneously proposed marriage to a girl. The rows of the matrix for the voting paradox can be used to show how she ranks them with respect to whatever three traits she considers most important, say intelligence, physical attractiveness, and income. Taken by pairs, the poor girl finds that she prefers A to B, B to C, and C to A. It is easy to see how similar conflicts can arise with respect to one's choice of a job, where to spend a vacation, and so on.

Paul R. Halmos once suggested a delightful interpretation of the matrix. Let A, B, and C stand for apple pie, blueberry pie, and cherry pie. A certain restaurant offers only two of them at any given meal. The rows show how a customer ranks the pies with respect to three properties, say taste, freshness, and size of slice. It is perfectly rational, says Halmos, for the customer to prefer apple pie to blueberry, blueberry to cherry, and cherry to apple. In his *Adventures of a Mathematician* (Scribner, 1976), Stanislaw Ulam speaks of having discovered the nontransitivity of such preferences when he was eight or nine, and of later realizing that it prevented one from ranking great mathematicians in a linear order of relative merit.

Experts differ on how often nontransitive orderings such as this one arise in daily life, but some recent studies in psychology and economics indicate that they are commoner than one might suppose. There are even reports of experiments with rats showing that under certain conditions the pairwise choices of individual rats are nontransitive. (*See* Warren S. McCulloch, "A Heterarchy of Values Determined by the Topology of Nervous Nets," *Bulletin of Mathematical Biophysics* 7, 1945, pp. 89–93.)

Similar paradoxes arise in round-robin tournaments between teams. Assume that nine tennis players are ranked in ability by the numbers 1 through 9, with the best player given the number 9 and the worst the number 1. The matrix in Figure 26 (right) is the familiar order-3 magic square. Let rows A, B, and C indicate how the nine players are divided into three teams with each row comprising a team. In round-robin tournaments between teams, where each member of one team plays once against each member of the others, assume that the stronger player always wins. It turns out that team A defeats B, B defeats C, and C defeats A, in each case by five games to four. It is impossible to say which team is the strongest. The same nontransitivity holds if columns D, E, and F of the matrix are the teams.

Many paradoxes of this type were jointly investigated by Leo Moser and J. W. Moon. Some of the Moser-Moon paradoxes underlie striking and little-known sucker bets. For example, let each row (or each column) of an order-3 magic-square matrix be a set of playing cards, say the ace, 6, and 8 of hearts for set A, the 3, 5, and 7 of spades for set B, and the 2, 4, and 9 of clubs for C (*see* Figure 27). Each set is randomized and placed face down on a table. The sucker is allowed to draw a card from any set, then you draw a card from a different

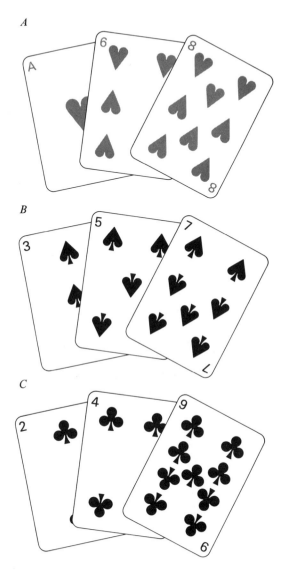

Figure 27 Nontransitive sucker bet based on magic square: $A \rightarrow B \rightarrow C \rightarrow A$

set. The high card wins. It is easy to prove that no matter what set the sucker draws from, you can pick a set that gives you winning odds of five to four. Set *A* beats *B*, *B* beats *C*, and *C* beats *A*. The victim may even be allowed to decide each time whether the high or the low card wins. If you play low card wins, simply pick the winning pile with respect to a nontransitive circle that goes the other way. A good way to play the game is to use sets of cards from three decks with backs of different colors. The packet of nine cards is shuffled each time, then separated by the backs into the three sets. The swindle is, of course, isomorphic with the tennis-tournament paradox.

Nontransitivity prevails in many other simple gambling games. (See Chapter 5 of my *Wheels, Life, and Other Mathematical Amusements*, for a description of a set of nontransitive dice.) In some cases, such as the top designed by Andrew Lenard (*see* Figure 28), the nontransitivity is easy to understand. The lower part of the top is fixed but the upper disk rotates. Each of two players chooses a different arrow, the top is spun (in either direction), and the person whose arrow points to the section with the highest number wins. *A* beats *B*, *B* beats *C* and *C* beats *A*, in each case with odds of two to one.

In a set of four bingo cards designed by Donald E. Knuth (*see* Figure 29), the nontransitivity is cleverly concealed. Two players each select a bingo card. Numbers from 1 through 6 are randomly drawn without replacement, as they are in standard bingo. If a called number is on a card, it is marked with a bean. The first player to complete a horizontal row wins. Here, of course, the numbers are just symbols; they can be replaced by any set of six different symbols. I leave it to the reader to work out the probabilities that show card *A* beats *B*, *B* beats *C*, *C* beats *D*, and *D* beats *A*. The game is transitive with three players, but the winning odds for the four possible triplets are surprising.

One of the most incredible of all nontransitive betting situations, discovered (appropriately) by a mathematician named Walter Penney, was given as a problem in the *Journal of Recreational Mathematics* (October 1969, p. 241). It is not well known, and most mathematicians simply cannot believe it when they first hear of it. It is certainly one of the finest of all sucker bets. It can be played

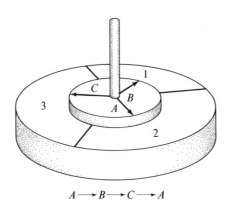

Figure 28 Nontransitive top

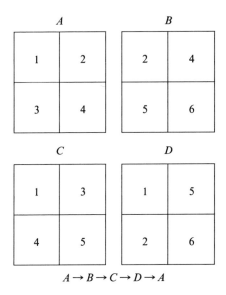

$$A \rightarrow B \rightarrow C \rightarrow D \rightarrow A$$

Figure 29 Nontransitive bingo cards

with a penny, or as a side bet on the reds and blacks of a roulette wheel, or in any situation in which two alternatives are randomized with equal odds. We shall assume that a penny is used. If it is flipped three times, there are eight equally probable outcomes: *HHH, HHT, HTH, HTT, THH, THT, TTH,* and *TTT.* One player selects one of these triplets, and the other player selects a different one. The penny is then flipped repeatedly until one of the chosen triplets appears as a run and wins the game. For example, if the chosen triplets are *HHT* and *THT* and the flips are *THHHT,* the last three flips show that *HHT* has won. In brief, the first triplet to appear as a run wins.

One is inclined to assume that one triplet is as likely to appear first as any other, but it takes only a moment to realize that this is not the case even with doublets. Consider the doublets *HH, HT, TH,* and *TT. HH* and *HT* are equally likely to appear first because, after the first *H* appears, it is just as likely to be followed by an *H* as by a *T.* The same reasoning shows that *TT* and *TH* are equal. Because of symmetry, *HH = TT* and *HT = TH. TH* beats *HH* with odds of three to one, however, and *HT* beats *TT* with the same probability. Consider *HT* and *TT. TT* is always preceded by *HT* except when *TT* appears on the first two flips. This happens in the long run only once in four times, and so the probability that *HT* beats *TT* is 3/4. Figure 30 shows the probability that *B,* the second player, will win for all pairs of doublets.

A / B	HH	HT	TH	TT
HH		1/2	1/4	1/2
HT	1/2		1/2	3/4
TH	3/4	1/2		1/2
TT	1/2	1/4	1/2	

Figure 30 Probabilities of *B* winning

When we turn to triplets, the situation becomes much more surprising. Since it does not matter which side of a coin is designated heads, we know that *HHH* = *TTT*, *TTH* = *HHT*, *HTH* = *THT*, and so on. When we examine the probabilities for unequal pairs, however, we discover that the game is not transitive. No matter what triplet the first player takes, the second player can select a better one. Figure 31 gives the probability that *B*, the second player, defeats *A* for all possible pairings. To find *B*'s best response to a triplet chosen by *A*, find *A*'s triplet at the top, go down the column until you reach a probability (shown in gray), then move left along the row to *B*'s triplet on the left.

Note that *B*'s probability of winning is, at the worst, 2/3 (or odds of two to one) and can go as high as 7/8 (or odds of seven to one). The seven-to-one odds are easy to comprehend. Consider *THH* and *HHH*. If *HHH* first appears anywhere except at the start, it must be preceded by a *T*, which means that *THH* has appeared earlier. *HHH* wins, therefore, only when it appears on the first three flips. Clearly this happens only once in eight flips.

Barry Wolk of the University of Manitoba has discovered a curious rule for determining the best triplet. Let *X* be the first triplet chosen. Convert it to a binary number by changing each *H* to zero and each *T* to 1. Divide the number by 2, round down the quotient to the nearest integer, multiply by 5, and add 4. Express the result in binary, then convert the last three digits back to *H* and *T*.

Nontransitivity holds for all higher *n*-tuplets. A chart supplied by Wolk gives the winning probabilities for *B* in all possible pairings of quadruplets (*see* Figure 32). Like the preceding two charts and charts for all higher *n*-tuplets, the matrix is symmetric about the center. The upper right quadrant is the lower left quadrant upside down, and the same holds for the upper left and lower right quadrants. The probabilities for *B*'s best responses to *A* are shown in gray.

In studying these figures, Wolk discovered another kind of anomaly as surprising as nontransitivity. It has to do with what are called waiting times. The waiting time for an *n*-tuplet is the average number of tosses, in the long run, until the specified *n*-tuplet appears. The longer you wait for a bus, the shorter becomes the expected waiting time. Pennies, however, have no memory, so that the waiting time for an *n*-tuplet is independent of all previous flips. The waiting time for *H* and *T* is 2. For doublets the waiting time is 4 for *HT* and *TH*, and 6 for *HH* and *TT*. For triplets the waiting times are 8 for *HHT*,

A / B	HHH	HHT	HTH	HTT	THH	THT	TTH	TTT
HHH		1/2	2/5	2/5	1/8	5/12	3/10	1/2
HHT	1/2		2/3	2/3	1/4	5/8	1/2	7/10
HTH	3/5	1/3		1/2	1/2	1/2	3/8	7/12
HTT	3/5	1/3	1/2		1/2	1/2	3/4	7/8
THH	7/8	3/4	1/2	1/2		1/2	1/3	3/5
THT	7/12	3/8	1/2	1/2	1/2		1/3	3/5
TTH	7/10	1/2	5/8	1/4	2/3	2/3		1/2
TTT	1/2	3/10	5/12	1/8	2/5	2/5	1/2	

Figure 31 Probabilities of *B* winning in a triplet game

A \ B	HHHH	HHHT	HHTH	HHTT	HTHH	HTHT	HTTH	HTTT	THHH	THHT	THTH	THTT	TTHH	TTHT	TTTH	TTTT
HHHH		1/2	2/5	2/5	3/10	5/12	4/11	4/11	1/16	3/8	3/8	3/8	1/4	3/8	7/22	1/2
HHHT	1/2		2/3	2/3	1/2	5/8	4/7	4/7	1/8	9/16	9/16	9/16	5/12	9/16	1/2	15/22
HHTH	3/5	1/3		1/2	3/5	5/7	1/2	1/2	5/12	5/12	9/16	9/16	5/14	1/2	7/16	5/8
HHTT	3/5	1/3	1/2		3/7	5/9	2/3	2/3	5/12	5/12	9/16	9/16	1/2	9/14	7/12	3/4
HTHH	7/10	1/2	2/5	4/7		1/2	1/2	1/2	7/12	7/12	5/14	1/2	7/16	7/16	7/16	5/8
HTHT	7/12	3/8	2/7	4/9	1/2		1/2	1/2	7/16	7/16	1/2	9/14	7/16	7/16	7/16	5/8
HTTH	7/11	3/7	1/2	1/3	1/2	1/2		1/2	1/2	1/2	9/16	5/12	7/12	7/12	7/16	5/8
HTTT	7/11	3/7	1/2	1/3	1/2	1/2	1/2		1/2	1/2	9/16	5/12	7/12	7/12	7/8	15/16
THHH	15/16	7/8	7/12	7/12	5/12	9/16	1/2	1/2		1/2	1/2	1/2	1/3	1/2	3/7	7/11
THHT	5/8	7/16	7/12	7/12	5/12	9/16	1/2	1/2	1/2		1/2	1/2	1/3	1/2	3/7	7/11
THTH	5/8	7/16	7/16	7/16	9/14	1/2	7/16	7/16	1/2	1/2		1/2	4/9	2/7	3/8	7/12
THTT	5/8	7/16	7/16	7/16	1/2	5/14	7/12	7/12	1/2	1/2	1/2		4/7	2/5	1/2	7/10
TTHH	3/4	7/12	9/14	1/2	9/16	9/16	5/12	5/12	2/3	2/3	5/9	3/7		1/2	1/3	3/5
TTHT	5/8	7/16	1/2	5/14	9/16	9/16	5/12	5/12	1/2	1/2	5/7	3/5	1/2		1/3	3/5
TTTH	15/22	1/2	9/16	5/12	9/16	9/16	9/16	1/8	4/7	4/7	5/8	1/2	2/3	2/3		1/2
TTTT	1/2	7/22	3/8	1/4	3/8	3/8	3/8	1/16	4/11	4/11	5/12	3/10	2/5	2/5	1/2	

Figure 32 Probabilities of *B* winning in a quadruplet game

HTT, THH, and *TTH;* 10 for *HTH* and *THT,* and 14 for *HHH* and *TTT.* None of this contradicts what we know about which triplet of a pair is likely to show first. With quadruplets, however, contradictions arise with six pairs. For example, *THTH* has a waiting time of 20 and *HTHH* a waiting time of 18. Yet, *THTH* is more likely to turn up before *HTHH* with a probability of 9/14, or well over one-half. In other words, an event that is less frequent in the long run is likely to happen before a more frequent event. There is no logical contradiction involved here, but it does show that "average waiting time" has peculiar properties.

There are many ways to calculate the probability that one *n*-tuplet will precede another. You can do it by summing infinite series, by drawing tree diagrams, by recursive techniques that produce sets of linear equations, and so on. One of the strangest and most efficient techniques was devised by John Horton Conway of the University of Cambridge. I have no idea why it works. It just cranks out the answer as if by magic, like so many of Conway's other algorithms.

The key to Conway's procedure is the calculation of four binary numbers that Conway calls leading numbers. Let A stand for the 7-tuplet $HHTHHHT$ and B for $THHTHHH$. We want to determine the probability of B beating A. To do this, write A above A, B above B, A above B, and B above A (see Figure 33). Above the top tuplet of each pair a binary number is constructed as follows. Consider the first pair, AA. Look at the first letter of the top tuplet and ask yourself if the seven letters, beginning with this first one, correspond exactly to the first seven letters of the tuplet below it. Obviously they do, and so we put a 1 above the first letter. Next, look at the second letter of the top tuplet and ask if the six letters, starting with this one, correspond to the *first six* letters of the tuplet below. Clearly they do not, and so we put zero above the second letter. Do the five letters starting with the third letter of the top tuplet correspond to the first five letters of the lower tuplet? No, and so this letter also gets zero. The fourth letter gets another zero. When we check the fifth letter of the top A, we see that HHT does correspond to the first three letters of the lower A, and so the fifth letter gets a 1. Letters six and seven each get zero. The "A leading A number," or AA, is 1000100, in which each 1 corresponds to a yes answer, each zero to a no. Translating 1000100 from binary to decimal gives us 68 as the leading number for AA.

Figure 33 shows the results of this procedure in calculating leading numbers AA, BB, AB, and BA. Whenever an n-tuplet is compared with itself, the first

$$1\ 0\ 0\ 0\ 1\ 0\ 0 = 68$$
$$A = HHTHHHT$$
$$A = HHTHHHT$$

$$0\ 0\ 0\ 0\ 0\ 0\ 1 = 1$$
$$A = HHTHHHT$$
$$B = THHTHHH$$

$$AA - AB : BB - BA$$
$$68 - 1 : 64 - 35$$
$$67 : 29$$

$$1\ 0\ 0\ 0\ 0\ 0\ 0 = 64$$
$$B = THHTHHH$$
$$B = THHTHHH$$

$$0\ 1\ 0\ 0\ 0\ 1\ 1 = 35$$
$$B = THHTHHH$$
$$A = HHTHHHT$$

Figure 33 John Horton Conway's algorithm for calculating odds of B's n-tuplet beating A's n-tuplet

digit of the leading number must, of course, be 1. When compared with a different tuplet, the first digit may or may not be 1.

The odds in favor of B beating A are given by the ratio $AA - AB : BB - BA$. In this case $68 - 1 : 64 - 35 = 67 : 29$. As an exercise, the reader can try calculating the odds in favor of THH beating HHH. The four leading numbers will be $AA = 7$, $BB = 4$, $AB = 0$, and $BA = 3$. Plugging these into the formula, $AA - AB : BB - BA$ gives odds of $7 - 0 : 4 - 3$, or seven to one, as expected. The algorithm works just as well on tuplets of unequal lengths, provided the smaller tuplet is not contained within the larger one. If, for example, $A = HH$ and $B = HHT$, A obviously wins with a probability of 1.

I conclude with a problem by David L. Silverman, who was the first to introduce the Penney paradox in the problems department that he then edited for the *Journal of Recreational Mathematics* (Vol. 2, October 1969, p. 241). The reader should have little difficulty solving it by Conway's algorithm. *TTHH* has a waiting time of 16 and *HHH* has a waiting time of 14. Which of these tuplets is most likely to appear first and with what probability?

ANSWERS

Which pattern of heads and tails, *TTHH* or *HHH*, is more likely to appear first as a run when a penny is repeatedly flipped? Applying John Horton Conway's algorithm, we find that *TTHH* is more likely to precede *HHH* with a probability of 7/12, or odds of seven to five. Some quadruplets beat some triplets with even greater odds. For example, *THHH* precedes *HHH* with a probability of 7/8, or odds of seven to one. This is easy to see. *HHH* must be preceded by a *T* unless it is the first triplet of the series. Of course, the probability of that is 1/8.

The waiting time for *TTHH* and for *THHH* is 16, compared with a waiting time of 14 for *HHH*. Both cases of the quadruplet versus the triplet, therefore, exhibit the paradox of a less likely event occurring before a more likely event with a probability exceeding 1/2.

ADDENDUM

Numerous readers discovered that Barry Wolk's rule for picking the best triplet B to beat triplet A is equivalent to putting in front of A the complement of its next-to-last symbol, then discarding the last symbol. More than half these correspondents found that the method also works for quadruplets except for the two in which H and T alternate throughout. In such cases the symbol put in front of A is the same as its next to last one.

Since October 1974, when this chapter first appeared in *Scientific American,* many papers have been published that prove Conway's algorithm and give procedures for picking the best *n*-tuplet for all values of *n*. Two important early articles are cited in the bibliography. The paper by Guibas and Odlyzko gives twenty-six references.

Readers David Sachs and Bryce Hurst each noted that Conway's "leading number," when an *n*-tuplet is compared with itself, automatically gives that tuplet's waiting time. Simply double the leading number.

Ancient Chinese philosophers (I am told) divided matter into five categories that form a nontransitive cycle: wood gives birth to fire, fire to earth, earth to metal, metal to water, and water to wood. Rudy Rucker's science-fiction story "Spacetime Donuts" (*Unearth,* Summer 1978) is based on a much more bizarre nontransitive theory. If you move down the scale of size, to several steps below electrons, you get back to the galaxies of the same universe we now occupy. Go up the scale several stages beyond our galactic clusters, and you are back to the elementary particles — not larger ones, but the very same particles that make our stars. The word "matter" loses all meaning.

The following letter was published in *Scientific American* (January 1975):

Sirs:

Martin Gardner's article on the paradoxical situations that arise from nontransitive relations may have helped me win a bet in Rome on the outcome of the Ali v. Foreman world heavyweight boxing title match in Zaïre on October 30.

Ali, though slower than in former years, and a 4–1 betting underdog, may have had a psychological and motivational advantage for that particular fight. But in addition, Gardner's mathematics might be relevant. Even though Foreman beat Frazier, who beat Ali, Ali could still beat Foreman because there may be a nontransitive relation between the three.

I ranked the three fighters against the criteria of speed, power, and technique (including psychological technique) as reported in the press, and spotted a nontransitive relation worth betting on:

	Ali	Frazier	Foreman
Speed	2	1	3
Power	3	2	1
Technique	1	3	2

Foreman's power and technique beat Frazier, but Ali's technique and speed beat Foreman. It was worth the bet. The future implications are, however, that Frazier can *still* beat Ali!

ANTHONY PIEL

Vaud, Switzerland

David Silverman (*Journal of Recreational Mathematics* 2, October 1969, p. 241) proposed a two-person game that he called "blind Penney-ante." It is based on the nontransitive triplets in a run of fair coin tosses. Each player *simultaneously* chooses a triplet without knowing his opponent's choice. The triplet that shows up first wins. What is a player's best strategy? This is not an easy problem. A full solution, based on an 8 × 8 game matrix, is given in *The College Mathematics Journal* as the answer to Problem 299 (January 1987, p. 74–76).

B I B L I O G R A P H Y

Nontransitive Voting and Tournament Paradoxes

Social Choice and Individual Values. Kenneth J. Arrow. Wiley, 1951.

Games and Decision. R. Duncan Luce and Howard Raiffa. Wiley, 1957.

The Theory of Committees and Elections. Duncan Black. Cambridge University Press, 1958.

"Voting and the Summation of Preferences: An Interpretive Bibliographical Review of Selected Developments During the Last Decade." William H. Riker, in *The American Political Science Review* 55, December 1961, pp. 900–911.

Topics on Tournaments. John W. Moon. Holt, Rinehart and Winston, 1968.

Theory of Voting. R. Farquharson. Yale University Press, 1969.

Paradoxes in Politics. Steven J. Brams. Free Press, 1976.

Arrow's Theorem: The Paradox of Social Choice, Alfred F. MacKay. Yale University Press, 1980.

Nontransitive Dice and Roulette Betting

"Nontransitive Dominance." Richard L. Tenney and Caxton C. Foster, in *Mathematics Magazine* 49, May 1976, pp. 115–120.

"A Coin Tossing Game." James C. Frauenthal and Andrew B. Miller, in *Mathematics Magazine* 53, September 1980, pp. 239–243.

"Lucifer at Las Vegas." Martin Gardner, Problem 27 in *Science Fiction Puzzle Tales.* Clarkson Potter, 1981.

"Nontransitive Dice and Other Probability Problems." Martin Gardner, in Chapter 5 of *Wheels, Life, and Other Mathematical Amusements.* W. H. Freeman and Company, 1983.

"Sheep Fleecing with Walter Funkenbusch." Ross Honsberger, in *Mathematical Gems III.* Mathematical Association of America, 1985.

Conway's Algorithm

"String Overlaps, Pattern Matching, and Nontransitive Games." L. J. Guibas and A. M. Odlyzko, in the *Journal of Combinatorial Theory* 30, March 1981, pp. 183–208.

"Coin Sequence Probabilities and Paradoxes." Stanley Collings, in the *Bulletin of the Institute of Mathematics and its Applications* 18, November/December 1982, pp. 227–232.

C H A P T E R

SIX

.

.

.

Combinatorial Card
Problems

There is no end to the making of mathematical tricks, puzzles, and other recreations that employ playing cards. In this chapter we look at some new card problems and games, with emphasis on how they lead into significant areas of combinatorial theory.

Consider the following combinatorial way of dramatizing an important number theorem. Remove all the cards of one suit (say, spades) from a deck and arrange them in serial order from ace to king. (The jack, queen, and king, respectively, represent 11, 12, and 13.) Place them face down in a row with the ace at the left. The following turning procedure is now applied, starting at the left at each step and proceeding to the right:

1. Turn over every card.
2. Turn over every second card. (Cards 2, 4, 6, 8, 10, and Q are turned face down.)
3. Turn over every third card.
4. Continue in this manner, turning every fourth card, every fifth card, and so on until you turn over only the last card.

Inspect the row. Note that all the cards except the ace, the 4, and the 9 are face down as shown in Figure 34. These values happen to be square numbers.

Is this an accident? Or is it an authentic hint of a general rule? A good classroom exercise is to prepare 100 small cards bearing numbers 1 through 100, stand them with their backs out in serial order on a blackboard ledge and apply the turning procedure. Sure enough, at the finish the only visible numbers will be the squares: 1, 4, 9, 16, 25, 36, 49, 64, 81, and 100. That is too large a sampling to be coincidental. The next step is to prove that no matter how large the deck, only squares survive the turning procedure.

A simple proof introduces one of the oldest and most fundamental of number theorems: A positive integer has an odd number of divisors (the divisors include 1 and the number itself), if and only if the number is a square. This is easy to see. Most divisors of a number come in pairs. Consider 72. The smallest divisor, 1, goes into the number 72 times, giving the pair 1 and 72. The next-larger divisor, 2, goes into the number 36 times, giving the pair 2 and 36. Similarly, $72 = 3 \times 24 = 4 \times 18 = 6 \times 12 = 8 \times 9$. The *only* divisor of a number that is not paired with a different number is a divisor that is a square root. Consequently, all nonsquares have an even number of divisors, and all squares have an odd number of divisors.

How does this apply to the row of cards? Consider the eight of spades in the first card-turning example. Since 8 is not a square, it has an even number of divisors: 1, 2, 4, 8. It will be turned four times: when you turn each card, each second card, each fourth card, and each eighth card. An even number of turns applied to a face-down card will leave that card face down. Since every nonsquare card will be turned an even number of times, it will be face down at the finish. The only cards that are turned an odd number of times and left face up are those with an odd number of divisors, namely the squares. Is there a better way to etch this basic number theorem in the memory of a high-school student than to have him witness such a demonstration?

Let us see how cards can be used for modeling a combinatorial problem that D. H. Lehmer once described as follows. Mr. Smith manages a motel. It consists of *n* rooms in a straight row. There is no vacancy. Smith is a psychologist who plans to study the effects of rearranging his guests in all possible ways. Every morning he gives them a new permutation. The weather is miserable, raining almost daily. To minimize his guests' discomfort, each daily rearrangement is made by exchanging the occupants of two adjoining rooms. Is there a simple algorithm that will run through all possible arrangements by switching adjacent occupants at each step?

The problem is easily modeled with cards. A row of spades, ace to king, corresponds to a thirteen-room motel. The number of permutations of *n* elements is factorial *n*. Our problem is to exchange two adjacent cards at each step and run through every possible permutation in just $(n! - 1)$ steps. (We

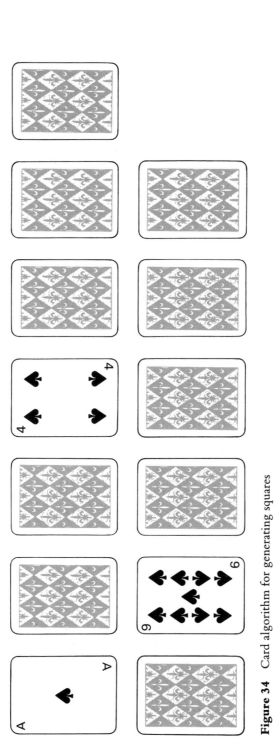

Figure 34 Card algorithm for generating squares

subtract 1 from $n!$ because we begin with one permutation on the table.) Such an algorithm has important applications in computer science. Many problems require a computer in order to run through all permutations of n elements, and if this can be done by exchanging adjacent pairs, there is a significant reduction in computer time.

It turns out that there is a simple, beautiful algorithm for doing this; it leads to the fastest-known way for a computer to permute n elements. Hugo Steinhaus, a Polish mathematician, was the first to discover it. It provides a solution for the abacus problem on page 49 of his *One Hundred Problems in Elementary Mathematics,* first published in Poland in 1958. In the early 1960's the procedure was independently rediscovered at almost the same time by H. F. Trotter and Selmer M. Johnson, each of whom published it separately.

Solving the problem for thirteen cards would require $13! - 1 = 6,227,020,799$ steps (you can see why a fast computer algorithm is desirable); hence let us start with smaller sets. It is easy to find a solution for three cards, but four cards present difficulties. Besides, we want not just a solution for a specified number of elements but a general method that will apply to any number.

With two cards ($n = 2$) the solution is trivial (*see* Figure 35A). Simply move the deuce from right to left. For $n = 3$ we list each of the preceding permutations three times: $12, 12, 12; 21, 21, 21$. The numeral 3 is now added to the list by a twisting procedure (*see* Figure 35B). The 3 starts at the right of the first permutation, weaves left through the series, pauses once at the left, then weaves back to end at the right of the final permutation. This generates the series $123, 132, 312, 321, 231$, and 213. If you start with an ace, deuce, and trey on the table and run through the series, you will see that each permutation is derived from the preceding one by switching two adjacent cards. For $n = 4$ each permutation of the series for $n = 3$ is repeated four times. (You always use n repetitions of the series for $n - 1$.) The numeral 4 is then added, weaving it left and right as before (*see* Figure 35C).

The algorithm can be defined by a nonrecursive formula, but the recursive procedure just explained is easier to understand. The trouble with defining a procedure for any n is that when the weaving card pauses at each side, the position of the pair to be switched varies in a curious way. The procedure given here is recursive, of course, because for each n, we must make use of the results obtained for ($n - 1$). It works for all higher n. For $n = 5$ we obtain $5! = 120$ permutations, beginning with $12345, 12354, 12534, 15234, 51234, 51243, 15243, \ldots$, and ending with 21345. Note that a switch of the first two numerals of the final permutation will give the first permutation. This holds for all n. The procedure is cyclic, restoring the original sequence in one more step. Note that after $n!/2$ steps the cards are in reverse order.

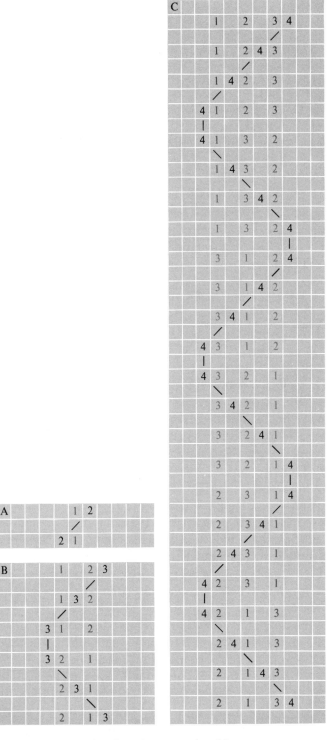

Figure 35 Recursive algorithm for solving motel problem

I am indebted to Donald E. Knuth for this means of displaying the recursive procedure, as well as for the algorithm's history. In the forthcoming fourth volume of his great series on *The Art of Computer Programming*, he will discuss the algorithm and show how its "inversion table" is equivalent to what is called a reflected Gray code with mixed bases. The problem is a special case of a more general "motel problem," which in turn is a special case of what Lehmer calls the "traveling-burglar problem."

A few years ago John Horton Conway of the University of Cambridge invented a series of card problems and games based on the technique of permuting a set of elements by reversing the order of subsets according to various rules. Take, for example, the thirteen spades from a deck, shuffle the packet and hold it face up in your left hand. Note the value of the uppermost card (we shall call it the top card) of the packet. Let us say it is a 9. Call the number out or to yourself, then with your right thumb slide nine cards off the packet one at a time into your right hand. This automatically reverses the order of the nine cards. Now put the nine-card packet on top of the cards in your left hand. A new card is face up on top. Note its value, call it out and repeat the same procedure. In other words, if n is the value of the top card, you always count off n cards, which reverses their order, and replace them face up on top. The game ends if an ace appears on top, because the ace produces a "one-loop" that consists in repeatedly counting off the ace and replacing it.

Must the game always end with the call of an ace? Yes, although it may take quite a while. It is impossible to get into a loop before the ace is called. If the game continues long enough without the call of an ace, a king might eventually be called. If this happens, however, the next reversal of cards puts the king on the bottom. Once the king is on the bottom, there is no way it can leave. As the game continues, the queen might eventually be called. If this happens, the queen goes to the twelfth position from the top, just above the king, and stays there. By mathematical induction the same thing must happen to the jack, then the 10 and so on (each card going to a position that corresponds with its value) until eventually, if not sooner, the ace is called and the game terminates. Indeed, each card can be called only once after the latest appearance of all higher cards.

The general form of the game involves one or more packets of n cards each. It is called k-swops if the kth card from the top is called. The called number gives the number of cards to be counted and replaced on top. It is called k-drops if the same procedure is followed, with the reversed set going to the bottom.

When there is only one packet and k equals 1 (the top card) and the counted cards go on top, Conway calls the game topswops. That is the game we have analyzed. The same game with the counted cards going to the bottom is called topdrops. In topdrops the top card is called, the cards are counted, and the reversed set is placed on the bottom. Topdrops is less interesting than top-swops. You begin at once with a loop that may be long or short and that may or may not contain an ace.

When there is one packet and k equals n (the bottom card), the game is botswops or botdrops, depending on whether the counted set goes on top or on the bottom. Botswops is boring. If you play with thirteen spades and the king is not on the bottom, you are immediately in a two-card loop. Suppose the bottom card is a 4. It stays there while you repeatedly reverse the four cards on top. If the king is at the bottom, it goes to the top, and you find yourself in a similar two-loop based on the new bottom card.

Botdrops (call the bottom card, count, and put the reversed set on the bottom) is more interesting. If you play it for a while, Conway writes, you might convince yourself that it always loops in a $KQKQKQ$ . . . sequence, but that is not always the case. On rare occasions other loops are possible. (Can you find one?) In this game, as in all the others, you start, of course, with a shuffled packet.

When the game is extended to two or more players, each with a packet, it becomes much harder to analyze. For instance, suppose two players have packets of thirteen cards each. One has spades, the other hearts. They play topswops as follows. Each shuffles his packet. Player A calls his top card, then B counts that number off his packet and replaces the reversed cards on top of his packet. B now calls *his* top card, A counts and replaces the reversed cards on top of his packet. This continues with players alternating calls.

It is a curious fact, reports Conway, that as soon as an ace is called, the calls go into a loop that starts with an ace, then a sequence, then an ace again (either the same ace or the other one), then the same sequence is repeated in reverse. For example, the first called ace might generate the following loop: $1-3-2-6-4-1-4-6-2-3-1$. Note that the sequence between the first two ace calls is the reverse of the sequence between the second and third ace calls. It is an unproved conjecture (or was when I last heard from Conway) that in two-player topswops an ace is always called. It is not known if the game can conclude in a loop without an ace, although it is known that if a loop includes an ace, it includes it just twice.

Let it not be supposed that these Conway card games are trivial. They deal with the theory of set permutations and not only may provide deep theorems but also may have a bearing on practical problems that arise in seemingly unrelated fields.

I conclude with three unusual combinatorial card problems. The first is extremely difficult, the second is easy but elegant, and the third is tricky.

1. A Langford problem. In Chapter 5 of my *Mathematical Magic Show* (Knopf, 1977), I discussed a combinatorial problem, first posed by C. Dudley Langford, that could be worked on with a set of cards containing doublets of values from 1 to n. When the problem is extended to triplets, the smallest value of n for which a solution is known is 9.

Here is the task. Remove from a deck all the cards of three suits that bear values of ace through 9. Try to arrange these twenty-seven cards in a single row to meet the following proviso. Between the first two cards of every value k there are exactly k cards, and between the second and third cards of every value k there also are exactly k cards. For instance, between the first and second 7's there must be just seven cards, not counting the two 7's. Similarly, seven cards separate the second and third 7's. The rule applies to each value from 1 through 9.

2. A Silverman problem. David L. Silverman is the inventor of this puzzle. Remove the spades and hearts from a deck. Put the spades face up in a row in serial order with the ace at the left and the king at the right. Place a heart card under each spade so that the sum of the two cards is a square number. Prove that the solution is unique.

3. A Ransom problem. This comes to me through the courtesy of Tom Ransom, a Canadian amateur magician and puzzle collector. Here is how Ransom has been showing his puzzle to magician friends. Five cards are placed in a row as shown in Figure 36. All card backs, Ransom states correctly, are either colored or black. Are all the cards with colored backs jokers?

The problem is not to answer the question but to determine the minimum number of cards that must be turned over in order to answer it. In other words, assuming any possible variation of the hidden card sides (each joker may have a black or a colored back, the card with the visible colored back may or may not be a joker and so on), how many cards must you turn over before you can answer the question: "Are all the cards with colored backs jokers?"

It is a confusing problem and one that calls for careful reasoning. There is a surprise in the solution that is closely related to an old joke about three professors on a train in Scotland. Through the window they see a black sheep.

Figure 36 Are all the cards with colored backs jokers?

"How interesting," says the astronomer. "All sheep in Scotland are black."

"A totally unwarranted inference," the physicist replies. "We can conclude only that *some* sheep in Scotland are black."

"Correction," says the logician. "At least one sheep in Scotland is black on at least one side."

ANSWERS

The three combinatorial problems are answered as follows:

1. Cards 1 through 9, of three suits, are to be arranged in a row so that for every card value k, just k cards are between the first and second cards of value k and between the second and third cards of value k. Not counting reversals, there are three solutions:

$$1, 8, 1, 9, 1, 5, 2, 6, 7, 2, 8, 5, 2, 9,$$
$$6, 4, 7, 5, 3, 8, 4, 6, 3, 9, 7, 4, 3$$

$$1, 9, 1, 2, 1, 8, 2, 4, 6, 2, 7, 9, 4, 5,$$
$$8, 6, 3, 4, 7, 5, 3, 9, 6, 8, 3, 5, 7$$

$$1, 9, 1, 6, 1, 8, 2, 5, 7, 2, 6, 9, 2, 5,$$
$$8, 4, 7, 6, 3, 5, 4, 9, 3, 8, 7, 4, 3$$

The third solution was found without computer aid in 1966 by Eugene Levine, now a mathematician at Adelphi University. It was published in 1968. Levine proves that a solution for triplets exists only when n, the highest value of a card, has a digital root of 1, 8, or 9 (that is, when n equals -1, 9, or 1, modulo 9), and that 9 is the smallest n that has a solution. Levine found solutions for the next higher cases of $n = 10, 17, 18$, and 19, and conjectured that there are solutions for all higher values meeting his proviso.

D. P. Roselle and T. C. Thomasson, Jr., writing on the problem in 1971, reported computer results that confirm there is no solution for $n = 8$, and came up with the same solution for $n = 9$ that Levine found. An exhaustive computer search for $n = 9$ and $n = 10$ was made by G. Baron, who reported his results at a conference on combinatorics held in Hungary in 1969. He found the three solutions for $n = 9$ and five solutions for $n = 10$. No solution has yet been found for this problem if there are more than three duplicates of each value.

2. The solution to David L. Silverman's problem of pairing each spade with a heart so that the pair-sum is square is: 1–8, 2–2, 3–K, 4–Q, 5–J, 6–10, 7–9, 8–1, 9–7, 10–6, J–5, Q–4, K–3. As Silverman observes, 9, 10, and J must be paired with 7, 6, and 5. This establishes six pairings. Since the 6's have been used, 3 pairs only with K. Since the 5's have been used, 4 pairs only with Q. The remaining three gaps are filled in only one way, proving the solution's uniqueness.

3. Tom Ransom's problem asked for the minimum number of cards, in a set of five, that must be turned over to answer the question: Are all colored-back cards jokers? Letter the cards in Figure 36 from A through E. Obviously, D must be turned to see if it is a joker, and E must be turned to see if it has a colored back. This gives four possibilities:

1. D is a joker, E has a black back.
2. D is a joker, E has a colored back.
3. D is not a joker, E has a black back.
4. D is not a joker, E has a colored back.

For cases 2, 3, and 4 the answer to the question is no. No more cards need be turned. For case 1 the answer is yes, but it takes more thinking to realize that turning the other three cards cannot contradict this answer. B is irrelevant because it has a black back. Seeing the back of either joker is also irrelevant. If a joker's back is black, it is not involved in the question. If it is colored, the answer is still yes. Most people staring at an actual row of cards have such an overwhelming desire to see the backs of the jokers that they usually answer: A, C, D, E.

One might conclude, therefore, that turning D and E is sufficient to answer the question. It is not! Recall the story about the cautious logician who observed a black sheep in Scotland and concluded that at least one sheep in Scotland is black on at least one side? When someone thinks he has solved the problem, Ransom turns over card B to reveal that its other side is a colored back! This, of course, contradicts a yes answer. The correct solution, therefore, is that card B as well as cards D and E must be turned.

Ransom has a second "kicker," suggested by his friend P. Howard Lyons. So that a person working on the problem will not forget the exact phrasing of the question ("Are all colored-back cards jokers?"), Ransom writes it on a file card which is placed above the row of cards. This card also must be turned to determine whether its back is colored or black!

ADDENDUM

The combinatorial playing-card problems produced memorable letters from readers who analyzed and extended the problems. Alan Hadsell and Stoddard Vandersteel together used a minicomputer to generalize David L. Silverman's problem. When the highest card value does not exceed 13, solutions exist only for $n = 3, 5, 8, 9, 10, 12$, and 13, and each solution is unique. From 14 through 31 all values of n have multiple solutions. They report that the number of solutions, beginning with $n = 14$, are 2, 4, 3, 2, 5, 15, 21, 66, 37, 51, 144, 263, 601, 1333, 2119, 2154, 2189, 3280,. . . .

Angeloo Papaikonomou, a bioengineer at the Free University of Amsterdam, after exploring the general problem for squares, turned to the general problem when cubes are substituted for squares. He was surprised to find that for every solvable n the solution is unique. The series of solvable n's begins 7, 19, 26, 37, 44, 56, 63, ... Papaikonomou found a simple recursive procedure that both gives this series and constructs the solutions.

Roland Silver of San Cristobal, N.M., exploring botdrops for cycles, discovered why the game usually terminates in a king–queen loop. There are no other 2-cycles and none of 3 and 4. Although 5-cycles exist, the probability of entering one is low. For example, if the face-up packet, reading up from the bottom, starts with 10, J, 2, 3 and has an ace on top, the packet is in a botdrop 5-loop.

Herbert S. Wilf of the University of Pennsylvania reported a delightful discovery about topswops that provides a proof of the game's finiteness. A card is in "natural position" if its value is the same as its position in the packet. For example, if the face-up packet, reading down from the top, is

$$7, 2, J, 8, 5, K, 6, A, 9, 10, 3, Q, 4$$

there are five cards (2, 5, 9, 10, Q) in natural position. If we take these values as powers of two, we can create what I shall call the Wilf number: $2^2 + 2^5 + 2^9 + 2^{10} + 2^{12} = 5668$. After any move in topdrops the Wilf number must increase.

"The reason that the number increases," Wilf writes, "is that the cards which were in natural position and which were too far down in the deck to be reached by the reversal operation will still be in natural position afterward. The fate of the cards which are involved in the reversal is less clear, except for one thing: the card which was on top before the move will be in natural position after the move, and its power of two is large enough to drown out any changes from cards above it. (A power of two is larger than the sum of all earlier powers of two by exactly one unit, a fact which is the basis of binary counting.)

"Since the numbers increase steadily but cannot exceed 16,382, it follows that the game must halt after at most that many moves. A slightly more careful study shows, in fact, that for a game with n cards, no more than 2^{n-1} moves can take place."

This raises an interesting unsolved question: What arrangement of the thirteen cards provides the longest possible game of topswops?

A simpler four-card version of Tom Ransom's puzzle, without Ransom's joke of using what magicians call a double-back card, has been the object of research by psychologists. It is considered, for example, in *The Psychology of Reasoning,* by Peter Wason and Philip Johnson-Laird.

Reaction to the disclosure that a card could be double-backed varied enormously. Jay Snyder began a letter: "Wait a minute! Hold it! Point of order! Not so fast! One moment please! Pull over, buddy." Stover sent a set of five taped-down cards to Wason at University College London for analysis. When Wason was later informed of what he would have found if he had pulled off the cards to check their undersides, he opened his next letter, "Mea culpa!"

Several readers found the problem's wording ambiguous as to the meaning of "minimum number of cards" to be turned. The problem is to specify a minimum set of cards that, when they are reversed, will in all possible cases guarantee a correct answer to the question: Are all red-backed cards jokers? (The problem is best presented with cards that may have either red or blue backs.) If we do not care whether or not our answer is correct, obviously we can "answer" the question by turning no cards. In some cases turning one card will guarantee a correct answer. What we want is the smallest set that will provide a correct answer for all cases.

James Weinrich has given me permission to reproduce here a letter he sent in December 1974:

> Dear Mr. Gardner,
>   Here are third and fourth "kickers" regarding Ransom's question, "Are all the colored-back cards jokers?"
>
> 3) The paperback *American Heritage Dictionary* (page 385) defines a joker as "an element in a situation that acts in an unexpected way." Clearly, a black-fronted card (B) that turns out to be red-backed is a joker by this definition. Accordingly: if B is not red-backed, it is irrelevant to the question; if it is red-backed, it is certainly an element acting in an unexpected way, and is thus a joker, but then

does not affect the answer as you explained for the other jokers. Thus it is not necessary to turn over B to know if all the red-backed cards are jokers.

4) The same dictionary (page 110) defines a card as "an amusing or eccentric person." If Ransom's card B is indeed red-backed, then he has played an amusing trick, and thus himself qualifies as a card. Accordingly, he could be turned over (or at least around) to see if he has acquired a sunburn on his back. One might reason that Ransom is a joker in any case, and thus it is not necessary to see if he has a red back to answer the question. However, Ransom is *not* a joker if the black-fronted card B is a standard one; in that case it is necessary to check Ransom's back.

Thus, the correct number of cards to turn is three or four: cards D and E in any case, card B (to see if Ransom is a joker), and Ransom (if B is not a joke card).

More or less sincerely yours,
JAMES D. WEINRICH

P.S.: Is Ransom Caucasian? If not, it can be assumed he has a colored back, and it would not be necessary to turn him over if B were a joke card. Note that in this case, his back might be both colored and black at the same time . . . .

B I B L I O G R A P H Y

The Motel Problem

"Algorithm 115." H. F. Trotter, in *Communications of the ACM* 5, 1962, pp. 434–435.

"Generation of Permutations by Adjacent Transposition." Selmer M. Johnson, in *Mathematics of Computation* 17, July 1963, pp. 282–285.

One Hundred Problems in Elementary Mathematics. Hugo Steinhaus. Basic, 1964. (See pp. 44–45, 49–50, 173–74.)

"Permutation by Adjacent Interchanges." D. H. Lehmer, in *The American Mathematical Monthly* 72, February 1965, pp. 36–46.

"Generating Permutations with Nondistinct Items." T. C. Hu and B. N. Tien, in *The American Mathematical Monthly* 83, October 1976, pp. 629–631.

Langford's Problem

"The Existence of Perfect 3-Sequences." Eugene Levine, in *The Fibonacci Quarterly* 6, November 1968, pp. 108–112.

"Über Verallgemeinerungen des Langford'schen Problems." G. Baron, in *Combinatorial Theory and Its Applications* 1. Paul Erdös et al., eds. Proceedings of the Conference in Balatonfüred, 1969. North Holland, 1970, pp. 81–92.

"On Generalized Langford Sequences." D. P. Roselle and T. C. Thomasson, Jr., in *Journal of Combinatorial Theory* 11, September 1971, pp. 196–199.

Ransom's Problem: Four-Card Versions

The Psychology of Reasoning. Peter Wason and Philip Johnson-Laird. Harvard University Press, 1972.

"The Psychology of Deceptive Problems." Peter Wason, in *New Scientist,* August 15, 1974, pp. 382–385.

The Mind's New Science: A History of the Cognitive Revolution, Chapter 13. Howard Gardner. Basic, 1985.

SEVEN

•
•
•

Melody-Making Machines

"Mathematics and music! The most
glaring possible opposites of human
thought! Yet connected, mutually
sustained!"

— HERMANN VON HELMHOLTZ,
Popular Scientific Lectures

There is a trivial sense in which any work of art is a combination of a finite
number of discrete elements. Not only that, the precise combination of the
elements can be expressed by a sequence of digits or, if you will, by one
enormous number.

Consider a poem. Assign distinct numbers to each letter of the alphabet, to
each punctuation symbol, and so on. A certain digit, say zero, can be used to
separate the numbers. It is obvious that one long string of digits can express the
poem. If the books of a vast library contain every possible combination of
words and punctuation marks, as they do in Jorge Luis Borges's famous story

"The Library of Babel," then somewhere in the collection is every poem ever written or that can be written. Imagine those poems coded as digital sequences and indexed. If one had enough time, billions on billions of years, one could locate any specified great poem. Are there algorithms by which one could find a great poem not yet written?

Consider a painting. Rule the canvas into a matrix of minute cells. The precise color of each cell is easily coded by a number. Scanning the cells yields a chain of numbers that expresses the painting. Since numbers do not decay, a painting can be re-created as long as the number sequence is preserved. Future computers will be able to reproduce a painting more like the original than the original itself, since after a few decades the original will have physically deteriorated to some extent. If a vast art museum contains every combination of colored cells for matrixes not exceeding a certain size, somewhere in that monstrous museum will hang every picture ever painted or that can be painted. Are there algorithms by which a computer could search a list of the museum's code numbers and identify a sequence for a great painting not yet painted?

Consider a symphony. It is a fantastically complex blend of discreteness and continuity; a violin or a slide trombone can move up and down the scale continuously, but a piano cannot produce quarter-tones. We know, however, from Fourier analysis that the entire sound of a symphony, from beginning to end, can be represented by a single curve on an oscilloscope. "This curve," wrote Sir James Jeans in *Science and Music* (Dover, 1968), "*is* the symphony — neither more nor less, and the symphony will sound noble or tawdry, musical or harsh, refined or vulgar, according to the quality of this curve." On a long-playing record a symphony is actually represented by one long space curve.

Because curves can be coded to any desired precision by numbers, a symphony, like a painting or a poem, can be quantized and expressed by a number chain. A vast library of computer tapes, recording all combinations of symphonic sounds, would contain every symphony ever written or that could be written. Are there algorithms by which a computer could scan the number sequences of such a library and pick out a great symphony not yet written?

Such procedures would, of course, be so stupendously complex that man may never come close to formulating them, but that is not the point. Do they exist in principle? Is it worthwhile to look for bits and pieces of them? Consider one of the humblest of such aesthetic tasks, the search for rules that govern the invention of a simple melody. Is there a procedure by which a person or a computer can compose a pleasing tune, using no more than a set of combinatorial rules?

If we restrict the tune to a finite length and a finite number of pure tones and rhythms, the number of possible melodies is finite. John Stuart Mill, in his autobiography, recalls that as a young man he was once "seriously tormented by the thought of the exhaustibility of musical compositions." Suppose our tune is made up of just ten notes chosen from the set of eight notes in a single octave. The number of melodies is the same as the number of ten-letter words that can be formed with eight distinct letters, allowing duplications. It is $8^{10} = 1,073,741,824$, and this without even considering varying rhythms which create, in effect, varying melodies. Most of these tunes will be dull (mi, mi, mi, mi, mi, mi, mi, mi, mi, mi for instance), but many will be extremely pleasing. Are there rules by which a computer or a person could pick out the pleasing combinations?

Attempts to formulate such rules and embody them in a mechanical device for composing tunes have a curious history that began in 1650 when Athanasius Kircher, a German Jesuit, published in Rome his *Musurgia universalis sive ars magna consoni et dissoni* (*see* Figure 37). Kircher was an ardent disciple of Ramón Lull, the Spanish medieval mystic whose *Ars magna* derived from the crazy notion that significant new knowledge could be obtained in almost every field simply by exploring all combinations of a small number of basic elements. It was natural that Kircher, who later wrote a 500-page elaboration of Lull's "great art," would view musical composition as a combinatorial problem. In his music book, he describes a Lullian technique of creating polyphony by sliding columns alongside one another, as with Napier's bones, and reading off rows to obtain various permutations and combinations. Like all of Kircher's huge tomes, the book is a fantastic mix of valuable information and total nonsense, illustrated with elaborate engravings of vocal cords, bones in the ears of various animals, birds and their songs, musical instruments, mechanical details of music boxes, water-operated organ pipes with animated figures of animals and people, and hundreds of other curious things.

The Lullian device described by Kircher was actually built, circa 1670, for the diarist Samuel Pepys, who owned a copy of Kircher's music book and much admired it. The original machine, called "musarithmica mirifica," is in the Pepys Museum at Pepys's alma mater, Magdalene College, Cambridge.

During the early eighteenth century many German music scholars became interested in mechanical methods of composition. Lorenz Christoph Mizler wrote a book in 1739 describing a system that produced figured bass for baroque ensemble music. In 1757 Bach's pupil, Johann Philipp Kirnberger, published in Berlin his *Ever-ready Composer of Polonaises and Minuets,* using a die for randomizing certain choices. In 1783 another book by Kirnberger extended his methods to symphonies and other forms of music.

Figure 37 Frontispiece of Kircher's *Musurgia universalis* (1650). The artist was J. Paul Schor. Here is how the picture is described in *Athanasius Kircher: A Renaissance Man and the Quest for Lost Knowledge,* by Joscelyn Godwin (Thames and Hudson, Ltd., 1979): "The symbol of the Trinity sheds its rays on the nine choirs of angels, who sing a 36-part canon (by Romano Micheli), and thence on the earth. The terrestrial sphere is shown encircled by the Zodiac and surmounted by Musica, who holds Apollo's lyre

Toward the close of the eighteenth century the practice of generating melodies with the aid of tables and randomizers such as dice or teetotums became a popular pastime. Maximilian Stadler, an Austrian composer, published in 1779 a set of musical bars and tables for producing minuets and trios with the help of dice. At about the same time, a London music publisher, Welcker, issued a "tabular system whereby any person, without the least knowledge of music, may compose ten thousand minuets in the most pleasing and correct manner." Similar anonymous works were falsely attributed to well-known composers such as C. P. E. Bach (son of Johann Sebastian Bach) and Joseph Haydn. The "Haydn" work, "Gioco filarmonico" ("Philharmonic Joke"), Naples, 1790, was discovered by the Glasgow mathematician Thomas H. O'Beirne to be a plagiarism. Its bars and tables are identical with Stadler's.

The most popular work explaining how a pair of dice can be used "to compose without the least knowledge of music" as many German waltzes as one pleases was first published in Amsterdam and in Berlin in 1792, a year after Mozart's death. The work was attributed to Mozart. Most Mozart scholars say it is spurious, although Mozart was fond of mathematical puzzles and did leave handwritten notes showing his interest in musical permutations. (The same pamphlet was issued in Bonn a year later, with a similar work, also attributed to Mozart, for dice composition of country dances. The contredanses pamphlet was reprinted in 1957 by Heuwekemeijer in Amsterdam.)

Mozart's *Musikalisches Würfelspiel,* as the waltz pamphlet is usually called, has been reprinted many times in many languages. In 1806 it appeared in London as *Mozart's Musical Game, Fitted in an Elegant Box, Showing by Easy System How to Compose an Unlimited Number of Waltzes, Rondos, Hornpipes and Reels.* In New York in 1941 the Hungarian composer and concert pianist Alexander Laszlo brought it out under the title *The Dice Composer,* orchestrating the music so that it could be played by chamber groups and orchestras. The system popped up again in West Germany in 1956 in a score published by B. Schott. Photocopies of Schott's charts and musical bars appear in the instruction booklet for *The Melody Dicer,* issued early in 1974 by Carousel Publishing

---

and the pan-pipes of Marsyas. In the landscape are seen dancing mermaids and satyrs, a shepherd demonstrating an echo, and Pegasus, the winged horse of the Muses. On the left is Pythagoras, the legendary father of musical theory. He points with one hand to his famous theorem, and with the other to the blacksmiths whose hammers, ringing on the anvil, first led him to discover the relation of tone to weight. On the right is a muse (Polymnia?) with a bird-perched on her head — possibly one of the nine daughters of Pierus, who for their presumption in attempting to rival the Muses were turned into birds. These figures are surrounded respectively by antique and modern instruments."

Corporation, Brighton, Mass. This boxed set also includes a pair of dice and blank sheets of music paper.

The "Mozart" system consists of a set of short measures numbered 1 through 176. The two dice are thrown sixteen times. With the aid of a chart listing eleven numbers in each of eight columns, the first eight throws determine the first eight bars of the waltz. A second chart is used for the second eight throws that complete the sixteen-bar piece. The charts are constructed so that the waltz opens with the tonic or keynote, modulates to the dominant, then finds its way back to the tonic on its final note. Because all bars listed in the eighth column of each chart are alike, the eleven choices (sums 2 through 12 on the dice) are available for only fourteen bars. This allows the system to produce 11^{14} waltzes, all with a distinct Mozartean flavor. The number is so large that any waltz you generate with the dice and actually play is almost certainly a waltz never heard before. If you fail to preserve it, it will be a waltz that will probably never be heard again.

The first commercial recording of "Mozart" dice pieces was made by O'Beirne. Both the randomizing of the bars and the actual playing of the melodies was done by Solidac, a small and slow experimental computer designed and built between 1959 and 1964 by the Glasgow firm of Barr and Stroud, where O'Beirne was then chief mathematician. It was the first computer built in Scotland. O'Beirne programmed Solidac to play the pieces in clarinetlike tones, and a long-playing recording of selected waltzes and contredanses was issued by Barr and Stroud in 1967. (This recording is no longer available.) O'Beirne is the author of an excellent book on mathematical recreations, *Puzzles and Paradoxes* (Oxford University Press, 1965). He has been of invaluable help in the preparation of this account.

Other methods of producing tunes mechanically were invented in the early 19th century. Antonio Calegari, an Italian composer, used two dice for composing pieces for the pianoforte and harp. His book on the system was published in Venice in 1801, and later in a French translation. *The Melographicon,* an anonymous and undated book issued in London about 1805, is subtitled: "A new musical work, by which an interminable number of melodies may be produced, and young people who have a taste for poetry enabled to set their verses to music for the voice and pianoforte, without the necessity of a scientific knowledge of the art." The book has four parts, each providing music for poetry with a certain meter and rhyme scheme. Dice are not used. One simply selects any bar from group *A,* any from group *B,* and so on to the last letter of the alphabet for that section.

A photograph of a boxed dice game appears in Plate 42 of *The Oxford Companion to Music,* but without mention of date, inventor, or place of publication. Apparently, it uses thirty-two dice, their sides marked to indicate tones, intervals, chords, modulations, and so on. There also are ivory men whose purpose, the caption reads, is "difficult to fathom."

In 1822 a machine called the Kaleidacousticon was advertised in a Boston music magazine, *The Euterpiad.* By shuffling cards it could compose 214 million waltzes. The Componium, a pipe organ that played its own compositions, was invented by M. Winkel of Amsterdam and created a sensation when it was exhibited in Paris in 1824. Listeners could not believe that the machine actually constructed the melodies it played. Scientists from the French Academy investigated.

"When this instrument has received a varied theme," their report stated, "which the inventor has had time to fix by a process of his own, it decomposes the variations of itself, and reproduces their different parts in all the orders of possible permutation . . . . None of the airs which it varies lasts above a minute; could it be supposed that but one of these airs was played without interruption, yet, the principle of variability which it possesses, it might, without ever resuming precisely the same combination, continue to play . . . during so immense a series of ages that, though figures might be brought to express them, common language could not."

The report, endorsed by physicist Jean Baptiste Biot, appeared in a British musical journal (*The Harmonicon* 2, 1824, pp. 40–41). Winkel's machine inspired a Vienna inventor, Baron J. Giuliani, to build a similar device, the construction of which is given in detail on pages 198–200 of the same volume.

In 1865, a composing system called the Quadrille Melodist, invented by J. Clinton, was advertised in *The Euterpiad.* By shuffling a set of composing cards, a pianist at a quadrille party could "keep the evening's pleasure going by means of a modest provision of 428,000,000 quadrilles."

Joseph Schillinger, a Columbia University mathematician who died in 1943, published his mathematical system of musical composition in a booklet, *Kaleidophone,* in 1940. George Gershwin is said to have used the system in writing *Porgy and Bess.* In 1940 Heitor Villa-Lobos, using the system, translated a silhouette of New York City's skyline into a piano composition (*see* Figure 38). *The Schillinger System of Musical Composition* is a two-volume work by L. Dowling and A. Shaw, published by Carl Fischer in 1941. A footnote on page 673 of Schillinger's eccentric opus *The Mathematical Basis of the Arts* (Philosophical Library, 1948) says that he left plans for music-composing machines, protected by patents, but nothing is said about their construction.

In the 1950s, information theory was applied to musical composition by J. R. Pierce and others. In a pioneering article "Information Theory and Melody," chemist Richard C. Pinkerton included a graph which he called the "banal tune-maker." By flipping a coin to determine paths along the network,

THE SKYLINE HAS ITS OWN MUSICAL PATTERN TRANSLATED
FROM SILHOUETTE TO MUSIC NOTES WITH THE HELP OF
THE SCHILLINGER SYSTEM OF MUSICAL COMPOSITION

Figure 38 New York skyline translated into music by Villa-Lobos, who used the Schillinger system

one can compose simple nursery tunes. Most of them are monotonous, but hardly more so, Pinkerton reminds us, than "A Tisket, a Tasket."

During the 1960s and early 1970s the proliferation of computers and the development of sophisticated electronic tone synthesizers opened a new era in machine composition of music. It is now possible to write computer programs that go far beyond the crude devices of earlier days. Suppose one wishes to compose a melody in imitation of one by Chopin. A computer analysis is made of all Chopin melodies so that the computer has in its memory a set of "transition probabilities." These give the probability that any set of one, two, three, or more notes in a Chopin melody is followed by any other note. Of course, one must also take into account the type of melody one wishes to compose, the rhythms, the position of each note within the melody, the overall pattern, and other things. In brief, the computer makes random choices within a specified general structure, but these choices are subject to rules and weighted by Chopin's transition preferences. The result is a "Markoff chain" melody, undistinguished but nevertheless sounding curiously like Chopin. The computer can quickly dash off several hundred such pieces, from which the most pleasing may be selected.

There is now a rapidly growing literature on computer composition, not only of music in traditional styles but also music that takes full advantage of the computer's ability to synthesize weird sounds that resemble none of the sounds made by familiar instruments. Microtones, strange timbres, unbelievably complex rhythms, and harmonics are no problem. The computer is a universal musical instrument. In principle, it can produce any kind of sound the human ear is capable of hearing. Moreover, a computer can be programmed to play one of its own compositions at the same time it is composing it.

How can we sum up? Computers certainly can compose mediocre music, frigid and forgettable, even though the music has the flavor of a great composer. No one, however, has yet found an algorithm for producing even a simple melody that will be as pleasing to most people of a culture as one of their traditional popular songs. We simply do not know what magic takes place inside the brain of a composer when he creates a superior tune. We do not even know to what extent a tune's merit is bound up with cultural conditioning or even with hereditary traits. About all that can be said is that a good melody is a mixture of predictable patterns and elements of surprise. What the proportions are and how the mixture is achieved, however, still eludes everybody, including composers.

O'Beirne has called my attention to how closely some systems of musical composition resemble the buzz-phrase generator (*see* Figure 39). This is a give-away of Honeywell Incorporated. Pick at random any four-digit number, such as 8751, then read off phrase 8 of module *A*, phrase 7 of module *B*, and so on. The result is a SIMP (Simplified Integrated Modular Prose) sentence. "Add a few more four-digit numbers," the instructions say, "to make a SIMP paragraph. After you have mastered the basic technique, you can realize the full potential of SIMP by arranging the modules in *DACB* order, *BACD* order, or *ADCB* order. In these advanced configurations, some additional commas may be required."

SIMP sounds very much like authentic technical prose, but on closer inspection one discovers that something is lacking. Computer-made melodies are perhaps less inane, closer to the random abstract art of a kaleidoscope, but still something essential (nobody knows what) is missing. Indeed, a good simple tune is much harder to compose than an orchestral piece in the extreme avant-garde manner, so loaded with randomness and dissonance that one hesitates to say, as Mark Twain (or was it Bill Nye?) said of Wagner's music: It is better than it sounds.

When a computer generates a melody that becomes as popular as (think of the title of your favorite song), you will know that a colossal breakthrough has been made. Will it ever occur? If so, when? Experts disagree on the answers as

SIMP table A	SIMP table B
1. In particular, 2. On the other hand, 3. However, 4. Similarly, 5. As a resultant implication, 6. In this regard, 7. Based on integral subsystem considerations, 8. For example, 9. Thus, 0. In respect to specific goals,	1. a large portion of the interface coordination communication. 2. a constant flow of effective information 3. the characterization of specific criteria 4. initiation of critical subsystem development 5. the fully integrated test program 6. the product configuration baseline 7. any associated supporting element 8. the incorporation of additional mission constraints 9. the independent functional principle 0. a primary interrelationship between subsystem and/or subsystem technologies

Figure 39 Honeywell's buzz-phrase generator for writing Simplified Integrated Modular Prose (SIMP)

much as they do on if and when a computer will write a great poem, paint a great picture, or play grand-master chess.

ADDENDUM

Carousel, the company that brought out Mozart's *The Melody Dicer*, later issued a similar set called *The Scott Joplin Melody Dicer*. Using the same system of dice and cards, one can compose endless rags of the Scott Joplin variety.

In 1977 a curious 284-page book titled *The Directory of Tunes and Musical Themes*, by Denys Parsons, was published in England by the mathematician G. Spencer Brown. Parsons discovered that almost every melody can be identified by a ridiculously simple method. Put down an asterisk for the first note. If the second note is higher, write down *U* for "up." If lower, write *D* for "down." If the same, use *R* for "repeat." Continue with succeeding notes until you have a sequence of up to sixteen letters. This is almost always sufficient to identify the melody. For example **UDDUUUU* is enough to key "White Christmas." The book lists alphabetically in two sections, popular and classical, some 15,000 different sequences followed by the title of the work, its composer, and the date.

SIMP table *C*	SIMP table *D*
1. must utilize and be functionally interwoven with	1. the sophisticated hardware
2. maximizes the probability of project success and minimizes the cost and time required for	2. the anticipated fourth generation equipment
3. adds explicit performance limits to	3. the subsystem compatibility testing
4. necessitates that urgent consideration be applied to	4. the structural design, based on system engineering concepts
5. requires considerable systems analysis and trade off studies to arrive at	5. the preliminary qualification limit
6. is further compounded, when taking into account	6. the evolution of specifications over a given time period
7. presents extremely interesting lenges to	7. the philosophy of commonality and standardization
8. recognizes the importance of other systems and the necessity for	8. the greater fight-worthiness concept
9. effects a significant implementation of	9. any discrete configuration mode
0. adds overriding performance constraints to	

The idea behind the buzz-phrase generator—random selection of words and phrases to create prose or poetry—is an old idea. *Rational Recreations,* a four-volume work by W. Hooper (the fourth edition was published in London in 1794), has a section in Volume 2 on how to use dice for composing Latin verse. The technique surely is much older. Similar randomizing is used in modern computer programs that generate "poems" and various kinds of imitation prose.

First-grade teachers, who call it "stringing," use the technique for teaching reading. Children are given a simple pattern sentence with blanks into which they insert words. The *New York Times Book Review* (June 4, 1978), printed Randolph Hogan's list of buzzwords that enable anyone to write impressive literary criticism. (Someone should do a similar list, perhaps already has, for art critics.) *Mad Magazine* (October 1974) featured Frank Jacobs's twelve columns of buzzwords and phrases for writing impeachment newspaper stories. Tom Koch (*Mad Magazine,* March 1982) gave a similar technique for stand-up comics. Jacobs returned in *Mad* (September 1982) with another twelve columns of words and phrases for writing the lyrics of country-western songs.

For Richard Voss's application of Benoit Mandelbrot's concept of fractals to the computer composition of music, see my *Scientific American* column for April 1978, and the references cited.

B   I   B   L   I   O   G   R   A   P   H   Y

"Information Theory and Melody." Richard C. Pinkerton, in *Scientific American,* February 1956, pp. 77–86.

"The Ars Magna of Ramon Lull." Martin Gardner, in *Logic Machines and Diagrams.* McGraw-Hill, 1958. Rev. ed. University of Chicago Press, 1982.

Experimental Music. L. A. Hiller, Jr., and L. M. Isaacson, McGraw-Hill, 1959.

"Computer Music." L. A. Hiller, Jr., in *Scientific American,* December 1959, pp. 109–120.

Music by Computers. Edited by Heinz Von Foerster and James W. Beauchamp. Wiley, 1969.

Cybernetic Serendipity: The Computer and the Arts. Jasia Reichardt, ed. Frederick A. Praeger, 1969.

"Composition Systems and Mechanisms." Percy A. Scholes, in *The Oxford Companion to Music,* John O. Ward, ed. Oxford University Press, 1970.

"From Mozart to the Bagpipe with a Small Computer." T. H. O'Beirne, in *Bulletin of the Institute of Mathematics and its Applications* 7, January 1971, pp. 3–8.

Athanasius Kircher. Joscelyn Godwin. Thomas and Hudson, 1979.

EIGHT

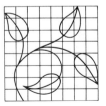

. . .

Anamorphic Art

Anamorphic art is a term unfamiliar to most people; indeed, it is unfamiliar to most artists. From the Greek *ana* (again) and *morphē* (form), it refers to realistic art so monstrously distorted by a projective transformation that it is difficult to recognize. The distortion can be "formed again" by viewing it on a slant or as a reflection in a suitable mirror. The mirror, called an anamorphoscope, is usually a polished cylinder or cone. The appearance of the undistorted reflection is so magical and surprising that few people seeing it for the first time fail to exclaim in wonder.

At this point the reader may want to pause to make the best cylindrical anamorphoscope he can in order to view some of the anamorphic art reproduced here and on the cover. For the best results, the cylinder should have a base that fits the circle in the picture. Acceptable results, however, can be obtained with a cylinder that has a larger or a smaller base. A small juice or soup can with the label removed and the residual glue on the can washed off may do the trick. A transparent cylindrical bottle with a cylinder of black paper inserted into it works fairly well. Chromium-plated tubing, available in some hardware stores for plumbing fixtures, is even better. Aluminum foil is not smooth and stiff enough, but silvered Mylar paper, taped around a cylinder of the right size, makes an excellent anamorphoscope. The reader is urged to get several square feet of this material (it is sold in art-supply and hobby stores), not only for cylindrical viewing, but also because it can be used, as will be explained, to make a conical anamorphoscope.

It was in the early Renaissance that European painters, who were just beginning to master perspective, became fascinated by the simplest kind of anamorphic art: stretched pictures that are seen correctly when viewed on a slant. The first known examples are in Leonardo da Vinci's notebooks; this is not surprising, because Leonardo was one of the earliest contributors to the geometry of perspective. Surfaces viewed on a slant are, of course, anamorphically distorted even though we are usually not aware of it. A door seen from a certain angle is a trapezoid, but our brain, conditioned by experience, perceives it as a tilted rectangle. When people who are not used to television see a television screen from the side, the images appear to be too skinny. The rest of us have learned to correct for this bias so well that squeezed images on television seem normal. When the Renaissance painters discovered how to transform flat shapes to give a depth illusion to the canvas, they discovered simultaneously how to do it in reverse. A picture stretched according to the same rules of perspective becomes a grotesque form.

Hans Holbein's painting *The Ambassadors* (1533) contains a famous example of anamorphic art (*see* Figure 40). You can see the stretched shape at the bottom of the painting normally by closing one eye and slanting the page away from you, with the lower left corner of the page pointing toward your open eye and about six inches from it. Another way to see the skull is to place the edge of a flat mirror about three inches from the lower left corner and to look into the mirror with both eyes while tipping it until the skull appears normal. Holbein's painting was probably intended to hang near the top of a stairway so that people going up would be startled by the skull.

Another slant picture is an old newspaper puzzle by Sam Loyd (*see* Figure 41). It has a concealed portrait of the mature George Washington. Can you find it? (A second puzzle consists in dividing the square Washington pie into six square pieces, not necessarily the same size.) Slant pictures of this kind occasionally appear in children's books and on advertising premiums. Sometimes, printing is stretched out so it can be read only by slanting the page. This technique is often used for the word STOP on streets so that the letters appear normal to a driver approaching the intersection.

The geometric technique for drawing slant pictures was explained in detail in the first major treatise on anamorphic art: *La Perspective Curieuse,* by Jean François Niceron (Paris, 1638). The picture is first ruled into square cells. The matrix is stretched to a trapezoid, then the artist copies the picture by filling in the trapezoidal cells, stretching the contents of each cell as accurately as he can to match its corresponding cell on the original picture. The finer the matrix, the more accurate the copy.

Figure 40 *The Ambassadors,* a slant anamorphic painting by Hans Holbein

The exact shape of the trapezoidal matrix depends on the position of the eye when it sees the shape as a normal square. The full three-dimensional structure is complex, but it turns out that there is a simple way to construct the trapezoid, given the desired position of the eye. Consider a square of side 8 ruled into 64 square-inch cells. We want to distort it so that it will appear normal when the eye is 25 units from the midpoint of the picture's top edge and seven units above the plane (*see* Figure 42). The construction is as follows. *XY* is the square's width. *FB,* on the perpendicular bisector of *XY,* is 25 units. *BE,* perpendicular to *FB,* is 7 units. Draw *XB, YB,* and *YE.* This determines *CD,* the trapezoid's bottom edge. Other lines from *B* and *E* to the unit marks on *XY*

Figure 41   A Sam Loyd puzzle with a hidden anamorphic picture

determine the lines to *XY* that complete the 64 trapezoidal cells. Neither *E* nor *B* indicates the eye's position. The eye is 7 units above the horizontal plane of the paper. A perpendicular dropped from the eye to the plane intersects *FB* at point *G*. The construction assumes that *G* is at least 8 units from *F*.

Another way to draw the picture is to close one eye, view the paper on a bias, and then draw the picture so that it looks normal. This is better than trying to do it with a mirror, because in the mirror your hand moves along one coordinate in a direction opposite to the way it moves on the actual sheet, and your reflexes find that hard to manage. A simple photographic method of making slant pictures is to project the picture (with an enlarger or a slide projector) onto enlarger paper so that the light strikes the paper at the angle intended for viewing.

Although there is no evidence that the ancient Greeks played with anamorphic pictures, they sometimes deformed the columns of temples to correct the distortion perceived by someone near the front of the building. For similar reasons, Renaissance painters occasionally deformed murals so that viewers looking up at them would see them with less distortion. Slant pictures were sometimes concealed in paintings or stretched along the side of a long corridor to be viewed from an entrance or an exit. Another popular practice was to put slant pictures inside boxes with a peephole at one end for viewing the picture at the proper angle.

Anamorphic paintings for cylindrical and conical mirrors were fashionable toys in both Europe and the Orient during the seventeenth and eighteenth centuries. They were usually done by anonymous artists and were sold with handsomely made anamorphoscopes. Occasionally the pictures carried messages of political protest; at times, they were pornographic or scatological. Several examples of erotic anamorphic pictures appear in *Chinese Erotic Art,* by Michel Beurdeley et al. (Charles E. Tuttle Co., 1969). This early anamorphic art is now a collector's item. In December 1973 a set of ten oil paintings, for both cylinders and cones, from eighteenth-century France was sold at a Sotheby Parke-Bernet auction in New York for $10,800, which is a bargain in view of today's prices. Herbert Tannenbaum, a New York art dealer, had found the paintings in 1939 in a curiosity shop in Amsterdam and had bought them without even knowing what they were. One of these paintings is shown in Figure 43.

Figure 44 produces the ten of hearts when viewed in a cylinder.

To make a conical anamorphoscope, cut a circular disk of the proper size from silvered Mylar, cut a radius, overlap the cut edges, and then glue or tape the overlapping edges in place. For the illustration in Figure 45, designed for conical viewing, the radius of the disk that makes a cone of the right propor-

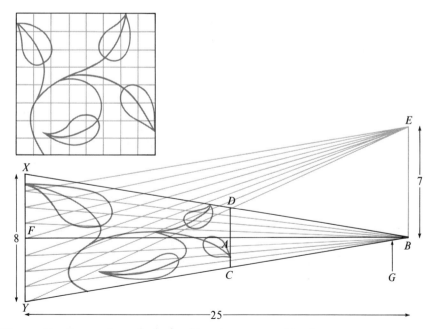

Figure 42 Geometric method of making slant anamorphic pictures

Figure 43 *With a Tender Little Song,* a cylindrical anamorphic painting from the
Tannenbaum Collection

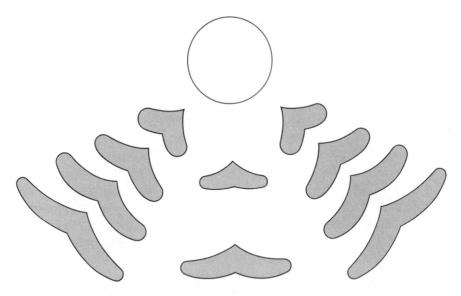

Figure 44   Cylindrical anamorphic picture

tions is about 1 inch. Adjust the overlap until the base of the cone fits the inner circle on the picture. Place the cone on this circle and view directly from above with one eye. The restored picture is small and circular, completely within the circumference of the cone. If you press on the apex of the cone with a fingertip or a paper clip, it will make the cone more rigid and produce a better picture. For ideal results, you need a solid conical mirror made with great accuracy.

As with slant pictures, there are three ways to deform a picture for cylindrical or conical viewing. The geometric procedures, reproduced from the Niceron treatise cited above, are shown in Figures 46 and 47. The methods for an exact construction of the distorted matrix are complicated; interested readers can do no better than to consult the Niceron book for details.

Note how the conical reflection literally turns a picture inside out. Point *A*, at the center of the picture in Figure 47, becomes the circumference of the distorted drawing, and the original circumference becomes the inner circle of the distortion.

When Salvador Dali made a set of erotic anamorphic paintings (prints were sold in Switzerland with a cylindrical anamorphoscope), he simply looked into the cylinder while he painted on the surface under it. That is not easy, because in the mirror your hand motion is reversed. You see what you are doing right side up, but your hand must paint the picture upside down.

One can make crude anamorphic photographic prints by wrapping a negative halfway around a transparent cylinder and by sending slanting light

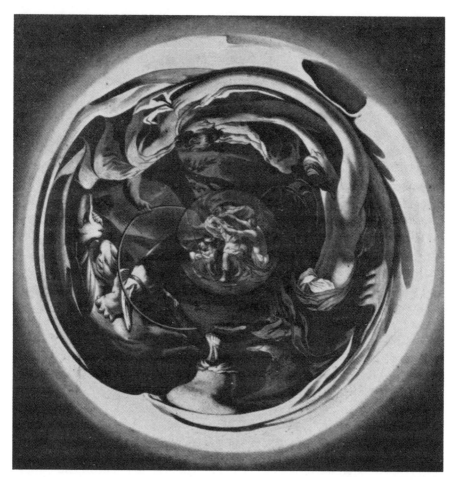

Figure 45 *Venus and Adonis,* a conical anamorphic painting. Photograph © Arnold Newman

through the cylinder from a point source outside and behind the cylinder to project the picture onto enlarging paper under the cylinder. More accurate prints are made by projecting the picture onto the side of an accurately made cylindrical mirror so that it reflects to the enlarging paper (*see* Figure 48). One should use an enlarger or a 35-millimeter slide projector with a diaphragm at the lens, and stop down the lens until the image is sharp on the easel. All light from the enlarger that does not fall directly on the cylinder should be blocked off by black cardboard with a rectangular hole. Conical prints are made in similar fashion. The mirror cone should be large (about 6 inches in diameter).

The picture is projected straight down on top of the cone through a circular hole in black cardboard to block off extraneous light.

The term anamorphic is used by photographers for any lens that stretches or compresses an image along one coordinate, as well as for the deformed images it produces. In 1953, Twentieth Century-Fox introduced the wide screen with its motion picture *The Robe*. Anamorphic lenses squeezed the wide image onto standard 35-mm. film, then anamorphic lenses in the projector stretched the image back to fit the wide screen. Most motion pictures today are taken and projected with sophisticated anamorphic lens systems. Similar systems adapt wide-screen motion pictures to videotape.

Psychologists who study perception have experimented with three-dimensional anamorphic models of chairs, tables, and other objects. The deformed models appear to be normal when seen from a certain angle. The Ames room is a radically distorted room that seems to be normal when viewed through a hole in the wall. A person in the room appears to grow or shrink when he or she moves from one part of the room to another (*see* "Experiments in Perception," by W. H. Ittelson and F. P. Kilpatrick; *Scientific American*, August

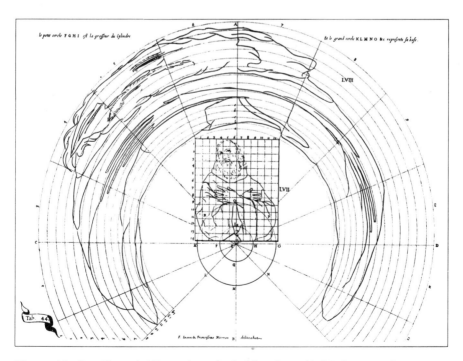

Figure 46 Jean François Niceron's method of drawing cylindrical anamorphic pictures

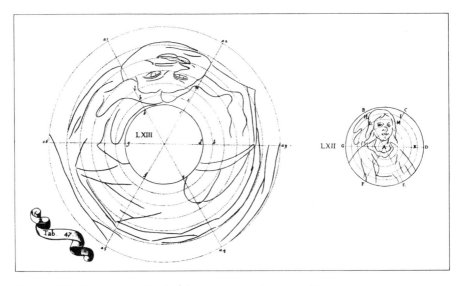

Figure 47 Niceron's method of drawing conical anamorphic pictures

1951). Seventeenth-century architects did not discover the Ames room, but they did play games with false perspective. The most startling example, which can still be seen in the Palazzo Spada in Rome, is an anamorphic arcade designed by Francesco Borromini about 1638. You seem to be looking down a long corridor at a large statue beyond the exit. Actually, the deformed corridor is only 28 feet long, and the statue is 3 feet high. The illusion was created by making the entrance 19 feet high and 10 feet wide and the exit only 8 feet high and 3⅓ feet wide (a trick, by the way, long familiar to designers of equipment for stage magicians).

There are many other forms of flat anamorphic art: pictures to be reflected in spheres, in properly placed *n*-sided pyramids and other polyhedrons, and pictures to be seen through various kinds of distorting lenses. The wavy mirrors in fun houses produce anamorphic images. What is a good caricature if not a complex set of anamorphic distortions that our mind sees as more like the person than the actual person? And there are extreme ways to transform a picture and to restore it again (a hologram, for instance, or the broadcasting of a television image), but the term *anamorphic* is best confined to coordinate transformations, particularly of the three types we have considered. Map makers do not use the term anamorphic, but the many ways in which the earth's surface is projected on the plane — cylindrically, conically, and otherwise — are coordinate transformations closely related to anamorphic art.

Botanists apply the word anamorphic to radical changes that certain plants undergo when they are grown in different environments. Zoologists have used the term for the evolutionary modifications of animal forms. D'Arcy Wentworth Thompson's classic work *On Growth and Form* (Cambridge University Press, 1961) has a chapter filled with diagrams showing animal species that differ from one another by anamorphic distortions so much like the types discussed here that if you view, say, one species of fish on a slant or in a cylindrical mirror, it becomes identical with another species. Similarly, a profile of the human skull, mildly anamorphosed, becomes the skull of a chimpanzee or a baboon.

The ability of our visual system to correct anamorphic distortion suggests that vision is concerned more with topologically invariant properties than

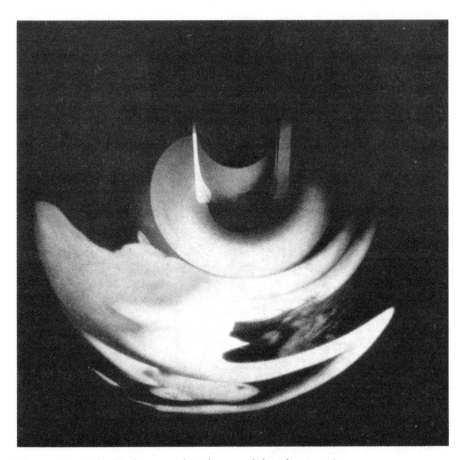

Figure 48 Cylindrical anamorphic photograph by Alan Fontaine

with Euclidean ones. The visual system not only utilizes the upside-down images on the two retinas in such a way as to provide a right-side-up impression of a three-dimensional world but also corrects for anamorphic distortion by irregular lenses and corneas. A person with marked astigmatism, fitted for the first time with glasses, perceives the world as being deformed because his brain is still correcting for the old distortions. It may take weeks before he sees the world normally again. Experiments have been conducted in which the subject wears special glasses that produce extreme topological transformations. After a few weeks and headaches, the world begins to look normal once again. When the glasses are removed, the world looks distorted, although fortunately only for a short time.

Hamlet advised some actors to hold the mirror up to nature. Is the mirror of a great play, novel, painting, or motion picture a distorting mirror or is it a magic anamorphoscope that gives pleasing form to an ugly, shapeless world? Are philosophical systems and religions, even the views of crazy little cults, anamorphic distortions of truth or are they, too, anamorphoscopes designed to give meaning to a meaningless reality? "It was to correct their anamorphosis of the Deity," wrote Thomas Jefferson, "that Jesus preached."

To the outsider, a system of beliefs appears to twist truth like a grotesque anamorphic painting. To the insider, who sees the world in the specially shaped mirror of his perceptual system, everything appears to be normal. Is there a metaphysical system that reflects truth like a flat, untipped mirror? Alas, every true believer is convinced his own anamorphoscope is precisely that.

ANSWERS

Although it has nothing to do with anamorphic art, Figure 41 contains a dissection puzzle that I did not answer in my column, but will do so here. It is one of Loyd's jokes. One assumes the square is to be divided along the lattice lines. Here is how Loyd answered the problem in his famous *Cyclopedia of 5,000 Puzzles* (Morningside Press, 1914). "The simplest way to cut a square into six squares is to mark it off into nine squares, then the largest one will be made up of four squares, and there will be five more little ones."

It is not only the simplest, it is the only solution. For a discussion of the general problem of dissecting squares into smaller squares, not necessarily alike, see the chapter on "Mrs. Perkins' Quilt" in my *Mathematical Carnival* (Knopf, 1975).

B I B L I O G R A P H Y

The Magic Mirror: An Antique Optical Toy. McLoughlin Brothers, ca. 1900. Dover reprint, 1979.

"Speaking of Pictures: Distorted Paintings Must Be Seen in Special Mirrors to Make Sense." *Life* 27, September 12 1949, pp. 18–19.

"The Grand Illusion." Fabrizio Clerici, in *Art News Annual* 23, 1954, pp. 98–180.

Anamorphoses: Ou Magie Artificielle Des Effets Merveilleux. Jurgis Baltrušaitis. Olivier Perrin Editeur, 1969.

"Anamorphic Art." Alan Fontaine. *Lithopinion* 7, Fall 1972, pp. 50–59.

Anamorphoses: Games of Perception and Illusion in Art. Joost Elffers, Fred Leeman, and Michael Schuyt. Abrams, 1976.

"Fun-Fair Illusions." Robert Hughes, in *Time,* October 4, 1976, pp. 92–93.

"Portrait of a Man Standing Before a Balustrade." Reprinting as a jigsaw puzzle of a 1630 Swedish painting, to be viewed in a cylindrical mirror. Abrams, 1977.

"Anamorphoscopes: A Visual Aid for Circle Inversion." Philip W. Kuchel, in *The Mathematical Gazette* 63, June 1979, pp. 82–89.

"Anamorphic Pictures." Jearl Walker, "The Amateur Scientist" column, in *Scientific American,* July 1981, pp. 176–187.

C H A P T E R

NINE

The Rubber Rope and
Other Problems

1. THE RUBBER ROPE

A worm is at one end of a rubber rope that can be stretched indefinitely (*see* Figure 49). Initially the rope is one kilometer long. The worm crawls along the rope toward the other end at a constant rate of one centimeter per second. At the end of each second the rope is instantly stretched another kilometer. Thus, after the first second, the worm has traveled one centimeter and the length of the rope has become two kilometers. After the second second, the worm has crawled another centimeter and the rope has become three kilometers long, and so on.

The stretching is uniform, like the stretching of a rubber band. Only the rope stretches. Units of length and time remain constant. We assume an ideal worm, of point size, that never dies, and an ideal rope that can stretch as long as needed.

Does the worm ever reach the end of the rope? If it does, estimate how long the trip will take and how long the rope will be. This delightful problem, which has the flavor of a Zeno paradox, was devised by Denys Wilquin of New Caledonia. It first appeared in December 1972 in Pierre Berloquin's lively puzzle column in the French monthly *Science et Vie*.

Figure 49   A worm crawling on a stretching rubber rope

2. THE SIGIL OF SCOTEIA

One of James Branch Cabell's finest novels, *The Cream of the Jest,* involves a series of hypnotic dreams that Felix Kennaston induces by staring at half of the Sigil of Scoteia. At the end of the novel, Kennaston discovers that the Sigil is nothing more than the broken top of a cold-cream jar designed by someone who "just made it up out of his head." It "is in no known alphabet. It blends meaningless curlicues and dots and circles with an irresponsible hand." The artist had sketched a crack across it "just to make it look ancient like."

A picture of the complete Sigil, which appears in most editions of the book, is reproduced in Figure 50. I remember puzzling over it many years ago and vainly trying to decode it. Designed by Cabell himself, it is not a cipher at all. Can the reader read it?

Cabell is still out of favor with most literary critics, but he has a loyal band of admirers who support the James Branch Cabell Society and its official organ, *Kalki.* It was in the sixth issue of *Kalki* (1968) that I learned the Sigil's secret.

3. INTEGER-CHOICE GAME

Two mathematicians are drinking beer. After each has been served a glass, they arrive at who pays for the round by simultaneously writing any positive integer on slips of paper. They compare the slips. Whoever wrote the larger integer has to pay for the round, unless the integer is larger by only 1. In that case the person who wrote the smaller integer pays for that round and the next as well. If both players have chosen the same integer, they play again.

To describe the game positively: The person who writes the smaller number scores a point, unless it is smaller by only 1. In that case the other player scores two points. For example, if the numbers were 12 and 20, 12 would win a point. If the numbers were 12 and 13, 13 would win two points.

The game is fair, but what strategy is best in the sense that no other strategy can beat it in the long run, and if any different strategy is followed, there is a counterstrategy to beat it?

The answer is surprising. I shall not have space for a proof, but I shall give the strategy and references to where the proof can be found. The game was devised by N. S. Mendelsohn and Irving Kaplansky, but it was Paul Halmos who called it to my attention.

4. THREE CIRCLES

Draw three nonoverlapping circles of three different sizes anywhere on a piece of paper. For each pair of circles draw their two common tangents. If you have not encountered this beautiful theorem before, you may be surprised to find that the intersections of the three pairs of tangents lie on a straight line (*see* Figure 51).

As one would expect, there are many ways the theorem can be proved by adding construction lines to the figure. *Popular Computing* reported in its issue

Figure 50 The Sigil of Scoteia

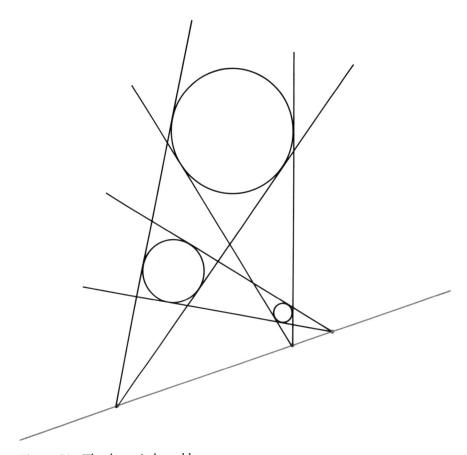

Figure 51 The three-circle problem

of December 1974, however, that the theorem lends itself to an elegant solution if one leaves the two-dimensional plane for a three-dimensional extension. Quoting from an earlier book in which they found the problem, the editors of the magazine report that when the theorem was shown to John Edson Sweet, a professor of engineering at Cornell University who died in 1916, he studied the picture for a moment and then said, "Yes, that's perfectly self-evident."

What was Professor Sweet's sweet solution?

5. THE MUTILATED SCORE SHEET

Figure 52 reproduces the score sheet of a chess game played in a German chess club in 1897. As you can see, a cigar or a cigarette has burned a hole in the sheet, so that Black's moves, which ended in Black's win on his fourth move, have been obliterated. Can you reconstruct the game?

I wish to thank Randolph W. Banner of Newport Beach, Calif., for passing along this amusing problem. He writes that he thinks he saw it in an English periodical published about 1920.

6. SELF-NUMBERS

D. R. Kaprekar is a mathematician, diminutive in body but large in brain and heart, who lives in India. For more than forty years he has been doing highly original work in recreational number theory, at times aided by grants from Indian universities. He contributes frequently to Indian mathematics journals, speaks at conferences, and has published some two dozen booklets written in broken English.

Kaprekar is best known outside India for his discovery, more than twenty years ago, of "Kaprekar's constant." Start with any four-digit number in which not all the digits are alike. Arrange the digits in descending order,

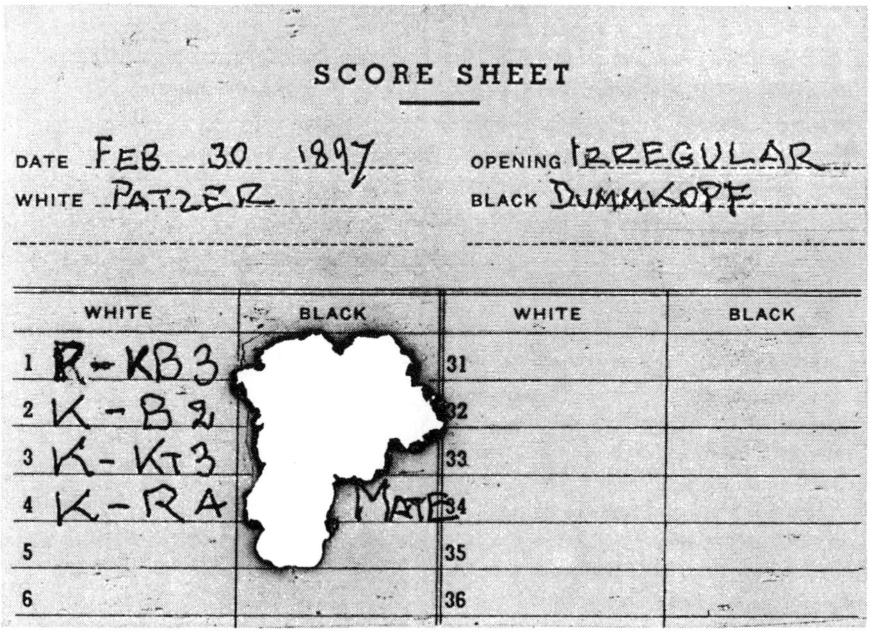

Figure 52 What were Black's moves?

reverse them to make a new number, and subtract the new number from the first number. If you keep repeating this process with the remainders, you will (in eight steps or fewer) arrive at Kaprekar's constant, 6174, which then generates itself. Zeros must be preserved. Thus, if you start with 2111 and subtract 1112, you get 0999. Rearranging the digits gives 9990 from which 0999 is taken, and so on.

Here we concern ourselves with a remarkable class of numbers called self-numbers, discovered by Kaprekar in 1949. He has written many pamphlets about them. They are virtually unknown outside India, although last year they turned up briefly (under another name) in an article in *The American Mathematical Monthly* (April 1974, p. 407). The article contains a proof that there is an infinity of self-numbers.

In explaining self-numbers, it is best to start with a basic procedure that Kaprekar calls digitadition. Select any positive integer and add to it the sum of its digits. Take 47, for example. The sum of 4 and 7 is 11, and 47 and 11 is 58. The new number, 58, is called a generated number. The original number, 47, is the generator. The process can be repeated endlessly, forming a digitadition series: 47, 58, 71, 79, 95, ...

No one has yet found a nonrecursive formula for the partial sum of a digitadition series, given its first and last terms, but there is a simple formula for the sum of all the digits in a digitadition series. Simply subtract the first number from the last and add the sum of the digits in the last number. "Is this not a wonderful new result?" Kaprekar asks in one of his booklets. "The Proof of all this rule is very easy and I have completely written it with me. But as soon as the proof is seen the charm of the whole process is lost, and so I do not wish to give it just now."

Can a generated number have more than one generator? Yes, but not until the number exceeds 100. The smallest such number (Kaprekar calls it a junction number) is 101. It has two generators: 91 and 100. The smallest junction number with three generators is 10,000,000,000,001. It is generated by 10,000,000,000,000, 9,999,999,999,901 and 9,999,999,999,892. The smallest number with four generators, discovered by Kaprekar on June 7, 1961, has twenty-five digits. It is 1 followed by twenty-one zeros and 102. Since then he has found what he conjectures to be the smallest numbers with five and six generators.

A self-number is simply a number that has no generator. In Kaprekar's words, "It is self born." There is an infinity of such numbers, but they are much scarcer than generated numbers. Below 100 there are thirteen: 1, 3, 5, 7, 9, 20, 31, 42, 53, 64, 75, 86, and 97. Self-numbers that are prime are called self-primes. The familiar cyclic number 142,857 is a self-number (multiply it by the digits 1 through 6 and you always get the same six digits in the same cyclic order). The numbers 11,111,111,111,111,111 and 3,333,333,333 are self-numbers. Previous years of this century that are self-numbers are 1906, 1917, 1919, 1930, 1941, 1952, 1963, and 1974.

Consider the powers of ten. The number 10 is generated by 5, 100 by 86, 1000 by 977, 10,000 by 9968, and 100,000 by 99,959. Why is a millionaire such an important man? Because, answers Kaprekar, 1,000,000 is a self-number! The next power of ten that is a self-number is 10^{16}.

No one has yet discovered a nonrecursive formula that generates all self-numbers, but Kaprekar has a simple algorithm by which any number can be tested to determine whether it is self-born or generated. Readers are asked to see if they can discover the procedure. If they can, it should be easy for them to answer this question: What is the next year after 1974 that is a self-number?

7. THE COLORED POKER CHIPS

What is the smallest number of poker chips that can be placed on a flat surface so that chips of three different colors are required to meet the condition that no two touching chips are the same color? As Figure 53 shows, the answer is obviously three.

Our problem is to determine the smallest number of chips, all of the same size, that can be placed flat on the plane so that chips of not three but four colors are necessary for meeting the same proviso.

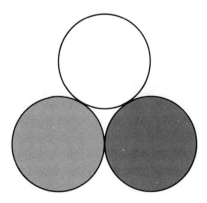

Figure 53   Map-coloring problem with poker chips

8. ROLLING CUBES

For this beautiful combinatorial puzzle, invented by John Harris of Santa Barbara, Calif., you must obtain eight unit cubes. On each cube, color one face and make the opposite face black. (Of course, you may distinguish the two faces in any other way you like.) Place the cubes in a shallow 3-by-3 box (or on a 3-by-3 matrix) with the middle cell vacant and all cubes black on top (*see* Figure 54).

A move consists in rolling a cube to an empty cell by tipping it over one of its four bottom edges so that it makes a quarter turn. The problem is to invert all eight cubes so their colored sides are up and the center cell is vacant as before. This is to be done in a minimum number of moves. For recording solutions, you can use *U, D, L, R* for roll up, down, left, and right, and start all solutions with *URD.*(Any other way of starting is symmetrically the same.)

ANSWERS

1. Regardless of the parameters (the initial length of the rubber rope, the worm's speed, and how much the rope stretches after each unit of time), the worm will reach the end of the rope in a finite time. This is also true if the stretching is a continuous process, at a steady rate, but the problem is easier to analyze if the stretching is done in discrete steps.

One is tempted to think one can see that the worm will make it. Since the rope expands uniformly, like a rubber band, the expansion is like looking at the rope through increasingly strong magnifying lenses. Because the worm is

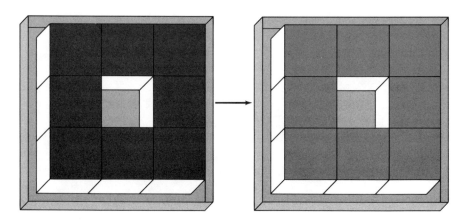

Figure 54   Rolling-cubes puzzle

always making progress, must it not eventually reach the end? Not necessarily. One can progress steadily toward a goal forever without ever reaching it. The worm's progress is measured by a series of decreasing fractions of the rope's length. The series could be infinite and yet converge at a point far short of the end of the rope. Indeed, such is the case if the rope stretches by doubling its length after each second.

The worm, however, does make it. There are 100,000 centimeters in a kilometer, so that at the end of the first second, the worm has traveled 1/100,000th of the rope's length. During the next second, the worm travels (from its previous spot after the stretching) a distance of 1/200,000th of the rope's length, after the rope has stretched to two kilometers. During the third second it covers 1/300,000th of the rope (now three kilometers), and so on. The worm's progress, expressed as fractional parts of the entire rope, is

$$\frac{1}{100,000}\left(1 + \tfrac{1}{2} + \tfrac{1}{3} + \tfrac{1}{4} + \cdots + \tfrac{1}{n}\right)$$

The series inside parentheses is the familiar harmonic one that diverges and therefore can have a sum as large as we please. The partial sum of the harmonic series is never an integer. As soon as it exceeds 100,000, however, the above expression will exceed 1, which means that the worm has reached the end of the rope. The number of terms, n, in this partial harmonic series will be the number of seconds that have elapsed. Since the worm moves one centimeter per second, n is also the final length of the rope in kilometers.

This enormous number, correct to within one minute, is

$$e^{100,000 - y} \pm 1$$

where y is Euler's constant. It gives a length of rope that vastly exceeds the diameter of the known universe, and a time that vastly exceeds present estimates of the age of the universe. (For a derivation of the formula see "Partial Sums of the Harmonic Series," by R. P. Boas, Jr., and J. W. Wrench, Jr., *American Mathematical Monthly* 78, October 1971, pp. 864–870.)

Some notion of the enormous length of the rope when the worm finally reaches the end can be gained from an observation by reader H. E. Rorschach. If the rope starts with a cross-sectional area of one square kilometer, it ends up as a single line of atoms, the space between each adjacent pair of atoms being many times the size of the known universe. The time lapse is comparably greater than the age of the universe.

2. As David M. Keller disclosed in his article "The Sigil of Scoteia" (*Kalki* 2, 1968), one simply turns the Sigil upside down. "Additional difficulties are found in the division of words at the ends of lines," Keller writes, "and in the substitution of odd characters for some of the letters." The Sigil reads: "James Branch Cabell made this book so that he who will may read the story of mans eternally unsatisfied hunger in search of beauty. Ettarre stays inaccessible always and her lovliness is his to look on only in his dreams. All men she must evade at the last and many are the ways of her elusion."

3. It is hard to believe, but the best strategy in the integer-choosing game is to limit one's choices of numbers to 1, 2, 3, 4, and 5. The selection is made at random, with the relative frequencies of 1/16 for numbers 1 and 5, 4/16 for number 3, and 5/16 for numbers 2 and 4. One could have in one's lap a spinner designed for picking numbers according to these frequencies.

For a proof of the strategy see "A Psychological Game," by N. S. Mendelsohn (*American Mathematical Monthly* 53, February 1946, pp. 86–88) and pages 212–215 of I. N. Herstein and I. Kaplansky's *Matters Mathematical* (Harper & Row, 1974).

Walter Stromquist, in a letter, proposed using a pair of dice as follows: "After a few beers, you cannot be expected to distinguish the fives and sixes, so that if either of these numbers appear on either die, you will have to roll again. Also, since you are really using only $\frac{2}{3}$ of each die, it is only natural to multiply the total by $\frac{2}{3}$ (dropping all fractions) before writing it down. For example, the largest number you can roll with two dice (without rolling again) is 8, so that the largest number you would ever choose is $\frac{2}{3}$ of that, or 5. Out of 16 plays, you should expect to choose with these frequencies: 1 once, 2 five times, 3 four times, 4 five times, 5 once." These are precisely the desired frequencies for playing the best strategy.

4. John Edson Sweet's solution to the three-circle theorem is given in the answer to Problem 62 in L. A. Graham's *Ingenious Mathematical Problems and Methods* (Dover, 1959). Instead of circles, suppose you are looking down on three unequal spheres. The tangent lines for each pair of balls are the edges of three cones into which the two balls fit snugly. The cones rest on the plane that supports the balls, and the apexes of the cones therefore lie on the plane.

Now imagine that a flat plate is placed on top of all three balls. Its underside is a second plane, tangent to all three balls, and tangent to all three cones. This second plane also will contain the three apexes of the cones. Because the apexes lie on both planes, they must lie on the intersection of the two planes, which of course is a straight line.

C. Stanley Ogilvy wrote to say he had included the three-circle problem in his book *Excursions in Geometry* (Oxford University Press, 1969). He too assumed that the proof by way of the three cones took care of all cases; but one day, after giving the problem to his class at Hamilton College, a student pointed out that the proof does not apply when a small sphere is between two larger ones. In such cases it is not possible for the two intersecting planes to be mutually tangent to all three spheres.

Many readers sent in other ways of proving the theorem. Bernard F. Burke, Richard I. Felver, Clyde E. Holvenstot, David B. Shear, and Radu Vero each proposed turning the drawing so that the line (on which the intersections of the tangent pairs lie) is horizontal and above the circles. The circles can now be viewed as being equal spheres inside three mutually intersecting pipes of identical circular cross section, seen in perspective. The tangent lines become the parallel sides of the three pipes. Since the pipes all must rest on a plane, their parallel sides seen in perspective will all have vanishing points on the horizon line.

It is not necessary that the circles be nonintersecting; indeed, the theorem can be stated in a more general way, in terms of "centers of similitude" instead of tangents, to hold for circles that lie entirely within one another. I am indebted to Donald Keeler for explaining this, as well as for pointing out that the theorem is known as "Monge's theorem" after the French mathematician and friend of Napoleon, Gaspard Monge, who gave it in a 1798 treatise. R. C. Archibald, in *The American Mathematical Monthly* (vol. 22, 1915, p. 65) traced the theorem back to the ancient Greeks (writes Keeler).

Daniel Sleator found that the theorem has an analog with four spheres in space. Each of the four triplets will have the apexes of its three cones on a straight line. Because these four lines intersect each other at six points, the four lines must be coplanar. Therefore, the vertexes of the six cones all lie on a plane. The theorem generalizes to all higher Euclidian dimensions. (See "Monge's Theorem in Many Dimensions," by Richard Walker, in *The Mathematical Gazette* 60, October 1976, pp. 185–188.)

Monge's theorem, for three circles on the plane, is mentioned in Herbert Spencer's autobiography. It is, writes Spencer, "a truth which I never contemplate without being struck by its beauty at the same time that it excites feelings of wonder and of awe: the fact that apparently unrelated circles should in every case be held together by this plexus of relations, seeming so utterly incomprehensible."

5.   The obliterated chess game is:

1. P–KB3	1. P–K4 or K3
2. K–B2	2. Q–B3
3. K–Kt.3	3. QxP (check)
4. K–R4	4. B–K2 (mate)

Several readers thought they had found a second solution. Black's first move is P–Q4. His second move is either P–KN3, P–KR4, or N–KB3. (Black's first two moves may be interchanged.) On his third move Black checks with Q–Q3, then presumably mates with Q–KB5. Unfortunately, the mate can be thwarted by White's fourth move, P–KN4.

6. D. R. Kaprekar's method of testing a number, N, to see if it is a self-number is as follows. Obtain N's digital root by adding its digits, then adding the digits of the result, and so on, until only one digit remains. If the digital root is odd, add 9 to it and divide by 2. If it is even, simply divide by 2. In either case call the result C.

Subtract C from N. Check the remainder to see if it generates N. If it does not, subtract 9 from the last result and check again. Continue subtracting 9's, each time checking the result to see if it generates N. If this fails to produce a generator of N in k steps, where k is the number of digits in N, then N is a self-number.

For example, we want to test the year 1975. Its digital root, 4, is even, so that we divide 4 by 2 to obtain C = 2. 1975 minus 2 is 1973, which fails to generate 1975. 1973 minus 9 is 1964. This also fails. But 1964 minus 9 is 1955, and 1955 plus the sum of its digits, 20, is 1975; therefore 1975 is a generated number. Since 1975 has four digits, we would have had only one more step to go to settle the matter. With this simple procedure, it does not take long to determine that the next self-year after 1974 is 1985. There will be only one more self-year in this century; 1996.

For progress on the problem of finding a nonrecursive formula for the sum of a digitadition series, see "The Sum of a Digitadition Series," by Kenneth B. Stolarsky (Proceedings of the American Mathematical Society 59, August 1976, pp. 1–5). Among his references on digitadition, the earliest is a 1906 French article.

7. It is easy to prove that the pattern of 11 circles (see Figure 55) requires at least four colors to ensure that no pair of touching circles are the same color. Assume that it can be done with three colors. Label circles 1, 2, and 3 with colors A, B, and C as shown. This determines the colors of 4, 5, and 6. We have a choice of two ways to color 7, but either way forces 11 to have the same color as 1, which it touches. Three colors are therefore not enough.

A number of readers sent proofs that 11 is the minimal number of circles. A proof by Allen J. Schwenk was published in The American Mathematical Monthly (vol. 83, June 1976, pp. 485–486) as a solution to part of question E 2527.

In three dimensions the least number of colors needed for any arrangement of identical spheres is not known, although it has been narrowed to 5, 6, 7, 8,

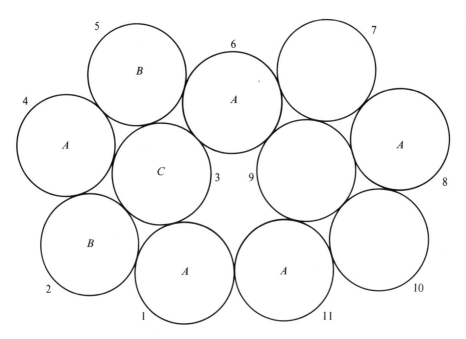

Figure 55   Solution to the poker-chip problem

or 9. For this and many other unsolved coloring problems, see "Coloring of Circles," by Brad Jackson and Gerhard Ringel (*The American Mathematical Monthly* 91, January 1984, pp. 42–49).

8.   In his article "Single Vacancy Rolling Cube Problems" *Journal of Recreational Mathematics* 7, Summer 1974, pp. 220–224), John Harris gave a thirty-eight-move solution to his rolling cube puzzle. About a dozen readers who wrote computer programs for the problem found a unique minimal-move solution in thirty-six moves. (Reversals, rotations, and reflections are not considered different.) The solution is:

URD LLD RRU LDL URD RUL DLU URD RUL DRD LUL DRU

Harris ends his article with a difficult problem that also involves eight cubes on an order-3 matrix. Color the cubes so that when they are on the matrix, with the center cell vacant, every exposed face is red and all hidden faces are uncolored. There will be just 24 red sides and 24 uncolored sides. The problem is to roll the cubes until they are back on the same eight cells, with the center cell vacant but with all red sides hidden and all visible sides uncolored.

Harris reported a solution in eighty-four moves, then later lowered it to seventy-four. In May 1981 I received a letter from Hikoe Enomoto, Kiyoshi Ishihata, and Satoru Kawai, all in the Department of Information Science, University of Tokyo. Their computer program produced the following minimal-move solution in seventy steps:

DRUUL   DDRUU

LDLDR   ULURD

RULDR   DLULD

RURDL   ULURD

RDLUR   DLURU

LDRUL   DLURD

RULDL   DRRUL

For two earlier rolling-cube puzzles invented by Harris, and a description of a board game based on rolling cubes, see Chapter 9 of my *Mathematical Carnival* (Knopf, 1975). For biographical information about Harris, see Steven Levy's *Hackers: Heroes of the Computer Revolution* (Doubleday, 1984).

TEN

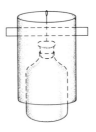

.
.
.

Six Sensational Discoveries

Has the steady rise of public interest in occultism and pseudoscience over the past ten years in the U.S. been something apart from public understanding of scientific knowledge? The two have interacted more strongly than most people realize. Important advances in science have been crowded out of newspapers, magazines, radio, and television to make room for reports on poltergeists, demon possession, psychic healing, prehistoric visits to the earth by astronauts from other worlds, the vanishing of ships and planes in the Bermuda Triangle, the emotional life of plants, the primal scream, and so on ad nauseam.

The effect is intensified by an increasing backlog of articles submitted to scientific journals. It is not unusual for several years to elapse between the acceptance of a scientific paper and its publication. In the meantime the author of an unpublished article about an important new discovery may keep his results secret for fear that a rival colleague might steal them and publish first.

As a public service, I shall comment briefly on six major discoveries of 1974 that for one reason or another were inadequately reported to both the scientific community and the public at large. The most sensational of last year's discoveries in pure mathematics was surely the finding of a counterexample to the notorious four-color-map conjecture. That theorem, as all readers of this book must know, is that four colors are both necessary and sufficient for coloring all planar maps so that no two regions with a common boundary are the same color. It is easy to construct maps that require only four colors, and topologists long ago proved that five colors are enough to color any map. Closing the gap, however, had eluded the greatest minds in mathematics. Most mathematicians have believed that the four-color theorem is true and that eventually it would be established. A few suggested it might be Gödel-undecidable. H. S. M. Coxeter, a geometer at the University of Toronto, stood almost alone in believing that the conjecture is false.

Coxeter's insight has now been vindicated. In November 1974 William McGregor, a graph theorist of Wappingers Falls, N.Y., constructed a map of 110 regions that cannot be colored with fewer than five colors (*see* Figure 56). McGregor's technical report will appear in 1978 in the *Journal of Combinatorial Theory,* Series B.

In number theory the most exciting discovery of the past year is that when the transcendental number *e* is raised to the power of π times $\sqrt{163}$, the result is an integer. The Indian mathematician Srinivasa Ramanujan had conjectured that *e* to the power of $\pi \sqrt{163}$ is integral in a note in the *Quarterly Journal of Pure and Applied Mathematics* (vol. 45, 1913–1914, p. 350). Working by hand, he found the value to be 262,537,412,640,768,743.999,999,999,999,. . . . The calculations were tedious, and he was unable to verify the next decimal digit. Modern computers extended the 9's much farther; indeed, a French program of 1972 went as far as two million 9's. Unfortunately, no one was able to prove that the sequence of 9's continues forever (which, of course, would make the number integral) or whether the number is irrational or an integral fraction.

In May 1974 John Brillo of the University of Arizona found an ingenious way of applying Euler's constant to the calculation and managed to prove that the number exactly equals 262,537,412,640,768,744. How the prime number 163 manages to convert the expression to an integer is not yet fully understood. Brillo's proof is scheduled to appear in a few years in *Mathematics of Computation*.

There were rumors late in 1974 that π would soon be calculated to six million decimal places. This may seen impressive to laymen, but it is a mere computer hiccup compared with the achievement of a special-purpose chess-playing computer built in 1973 by the Artificial Intelligence Laboratory at the Massachusetts Institute of Technology. Richard Pinkleaf, who designed the computer with the help of ex-world-chess-champion Mikhail Botvinnik of the U.S.S.R., calls his machine MacHic because it so often plays as if it were intoxicated.

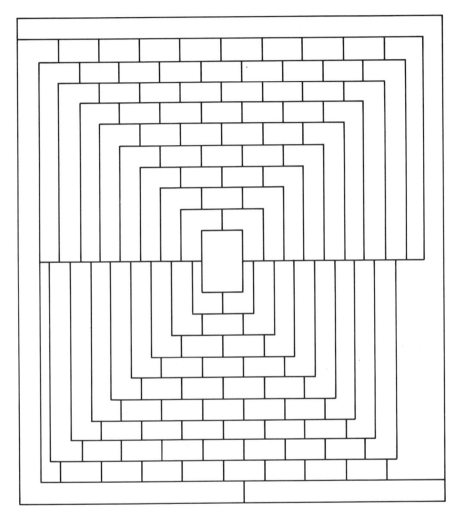

Figure 56 The four-color-map theorem is exploded

Unlike most chess-playing programs, MacHic is a learning machine that profits from mistakes, keeping a record of all games in its memory and thus steadily improving. Early in 1974 Pinkleaf started MacHic playing against itself, taking both sides and completing a game on an average of every 1.5 seconds. The machine ran steadily for about seven months.

At the end of the run, MacHic announced an extraordinary result. It had established, with a high degree of probability, that pawn to king's rook 4 is a win for White. This was quite unexpected because such an opening move has traditionally been regarded as poor. MacHic could not, of course, make an exhaustive analysis of all possible replies. In constructing a "game tree" for the opening, however, MacHic extended every branch of the tree to a position that any chess master would unhesitatingly judge to be so hopeless for Black that Black should at once resign.

Pinkleaf has been under enormous pressure from world chess leaders to destroy MacHic and suppress all records of its analysis. The Russians are particularly concerned. I am told by one reliable source that a meeting between Kissinger and Brezhnev will take place in June, at which the impact on world chess of MacHic's discovery will be discussed.

Bobby Fischer reportedly said that he had developed an impregnable defense against P-KR4 at the age of eleven. He has offered to play it against MacHic, provided that arrangements can be made for the computer to play silently and provided that he (Fischer) is guaranteed a win-or-lose payment of $25 million.

The reaction of chess grand masters to MacHic's discovery was mild compared with the shock waves generated among leading physicists by last year's discovery that the special theory of relativity contains a logical flaw. The crucial "thought experiment" is easily described. Imagine a meterstick traveling through space like a rocket, on a straight line colinear with the stick. A plate with a circular hole one meter in diameter is parallel to the stick's path and moving perpendicularly to it (*see* Figure 57). We idealize the experiment by assuming that both the plate and the meterstick have zero thickness. The

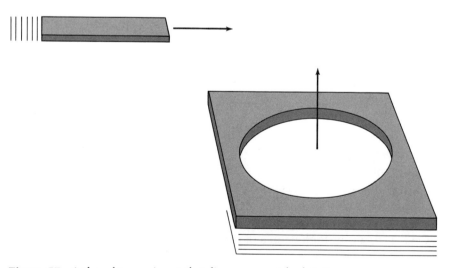

Figure 57 A thought experiment that disproves special relativity

two objects are on a precise collision course. At the same instant, the center of the meter stick and the center of the hole will coincide.

Assume that the plate is the fixed inertial frame of reference and the meterstick is moving so fast that it is Lorentz-contracted by a factor of 10. In this inertial frame the stick has a length of 10 centimeters. As a result it will pass easily through the hole in the rapidly rising plate. (The speed of the rising plate is immaterial.)

Now consider the situation from the standpoint of the meterstick's inertial frame. The plate is moving in the opposite horizontal direction, and so now it is the hole that is Lorentz-contracted, along its diameter parallel to the stick, to 10 centimeters. There is no way the 10 centimeter by 1 meter elliptical hole can move up past the meterstick without a collision. The two situations are not equivalent, and thus a fundamental assumption of special relativity is violated.

Physicists have long realized that the general theory of relativity is weakly confirmed, but the special theory had been confirmed in so many ways that its sudden collapse came as a great surprise. Humbert Pringle, the British physicist who discovered the fatal *Gedankenexperiment,* reported it in a short note last summer in *Reviews of Modern Physics,* but the full impact of all of this has not yet reached the general public.

When facsimiles of two lost notebooks of Leonardo da Vinci's were published in 1974 by McGraw-Hill, they were widely reviewed. The public learned of many hitherto unknown inventions made by Leonardo: a system of ball bearings surrounding a conical pivot (thought to have been first devised by Sperry Gyroscope in the 1920s), a worm screw credited to an eighteenth-century clockmaker, and dozens of other devices, including a bicycle with a chain drive.

In view of the publicity given the McGraw-Hill volumes, it is hard to understand why the media failed to report in December, 1974 on the discovery of a drawing that had been missing from the first notebook. This notebook, known as *Codex Madrid I* (it had been found ten years earlier in the National Library in Madrid), is a systematic treatise of 382 pages on theoretical and applied mechanics (*see* "Leonardo on Bearings and Gears," by Ladislao Reti; *Scientific American,* February 1971). There had been much speculation on the nature of the missing page. Augusto Macaroni of the Catholic University of Milan observed that the sketch was in a section on hydraulic devices, and he speculated that it dealt with some type of flushing mechanism.

The missing page was found shortly before Christmas by Ramón Paz y Bicuspid, head of the manuscript division of the Madrid Library. It was Bicuspid who had originally found the two lost notebooks. The missing page had been torn from the manuscript and inserted in a fifteenth-century treatise on the Renaissance art of perfume making. Figure 58 reproduces a photocopy of the original drawing. As the reader can see at once, Professor Macaroni was on target. The drawing establishes Leonardo as the first inventor of the valve flush toilet.

It had long been known that Leonardo had invented a folding toilet seat and had proposed a water closet with continuously running water in channels inside walls, a ventilating shaft to the roof, and suspended weights to make sure the entrance door closed. Until now, however, the first valve flush toilet has always been credited to Sir John Harington, a godson of Queen Elizabeth. Harington described it amusingly in his book *The Metamorphosis of Ajax* (1596) a cloacal satire that got him banished from the court. Although his "Ajax" actually was built at Kelston near Bath, it was not until 200 years later that it came into general use.

The first English patent for a valve flush toilet was granted in 1775 to Alexander Cummings, a watchmaker. Modern mechanisms, in which a ball float and automatic cutoff stopper limit the amount of water released with

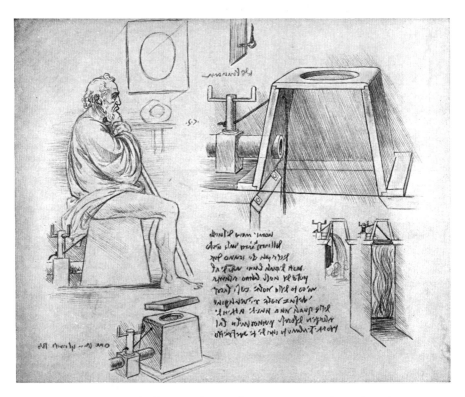

Figure 58 Lost Leonardo da Vinci drawing [courtesy N.Y. Public Library]

each flush, date from the early nineteenth-century patents of Thomas Crapper, a British manufacturer of plumbing fixtures who died in 1910. (See *Clean and Decent: The Fascinating History of the Bathroom and Water Closet,* by Lawrence Wright, Routledge and Kegan Paul, 1960, and *Flushed with Pride: The Story of Thomas Crapper,* by Wallace Reyburn, Prentice-Hall, 1971.)

Although hundreds of books on parapsychology spewed forth from reputable publishing houses in 1974, not one reported the most sensational psi discovery of the century: a simple motor that runs on psi energy. It was constructed in 1973 by Robert Ripoff, the noted Prague parapsychologist and founder of the International Institute for the Investigation of Mammalian Auras. When Henrietta Birdbrain, an American expert on Kirlian photography, visited Prague early last year, Dr. Ripoff taught her how to make his psychic motor. Ms. Birdbrain demonstrated the device many times in her lectures, but as far as I am aware the only published report on it appeared in the Boston monthly newspaper *East West Journal* (May 1974, p. 21).

Readers are urged to construct and test a model of the motor. The first step is to cut a three-by-seven-inch rectangle from a good grade of bond paper. Make a tiny slot in the paper at the spot shown (*see* Figure 59). The slot must be $\frac{3}{8}$ inch long and exactly in the center of the strip, $\frac{1}{8}$ inch from the top edge. Bend the paper into a cylinder, overlapping the ends $\frac{5}{16}$ inch, and glue the ends together. Cut a second slot in the center of the overlap, directly opposite the preceding one. It must be the same size and the same distance from the top.

From a file card or a piece of pasteboard of similar weight, cut a strip $\frac{3}{8}$ inch by three inches. Insert a fine, sharp-pointed needle twice through the center of the strip, as shown in step 3. The point of the needle should be no more than $\frac{1}{4}$ inch below the bottom edge of the strip. Push the ends of the strip through the cylinder's two slots, as shown in step 4, taking care not to bend the strip. The final step is to balance the needle on top of a narrow bottle at least four inches high (step 5). It is essential that the top of the bottle be either glass (preferable) or very hard smooth plastic.

Adjust the strip in the slots until the cylinder hangs perfectly straight, its side the same distance from the bottle all around. With scissors snip the ends of the strip so that each end projects $\frac{1}{4}$ inch on each side.

Place the little motor on a copy of the Bible or the *I Ching,* with the book's spine running due north and south. Sit in front of the motor, facing north. Hold either hand, cupped as shown in Figure 60, as close to the cylinder as you can without actually touching it. You must be in a quiet room, where the air is still. Make your mind blanker than usual and focus your mental energy on the motor. Strongly will it to rotate either clockwise or counterclockwise. Be patient. It is normally at least one full minute before the psi energy from your aura takes effect. When it does, the cylinder will start to rotate slowly.

STEP 1

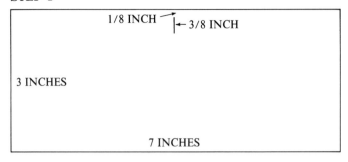

1/8 INCH — |← 3/8 INCH

3 INCHES

7 INCHES

STEP 2   **STEP 3**

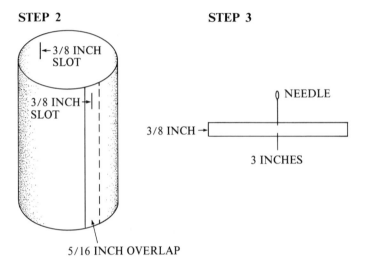

|← 3/8 INCH
   SLOT

3/8 INCH →|
SLOT

5/16 INCH OVERLAP

3/8 INCH →

NEEDLE

3 INCHES

STEP 4   **STEP 5**

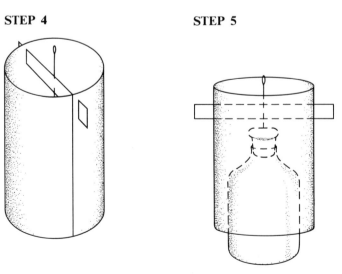

Figure 59 A psychic motor is made

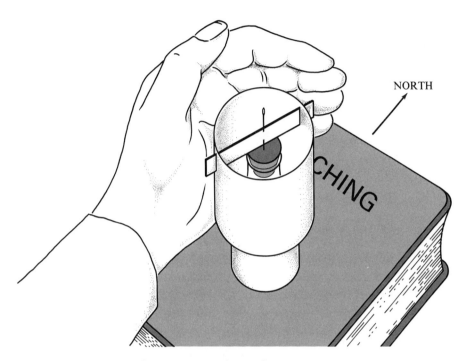

Figure 60 How to apply psi energy to the psychic motor

Some people, of course, have a stronger psi field than others. A lot depends on your mental state. At times the motor refuses to turn. At other times it begins to turn almost as soon as you start concentrating. Experiments show that for most people it is easier to make the motor rotate counterclockwise with psi energy from your right hand and clockwise with the energy from your left hand. At times negative psi takes over, and the motor turns in the direction opposite to the direction being willed. As Dr. J. B. Rhine has taught us, psi effects are elusive, skittish, and unpredictable.

The motor is currently under extensive investigation at numerous parapsychology laboratories around the world. Russian experts are convinced the energy that turns the motor is the same as the psychokinetic energy that enables the Israeli psychic Uri Geller to bend silverware, the Russian "sensitive" Ninel Kulagina to levitate table-tennis balls, and the Brooklyn psychic Dean Kraft to make pieces of candy leap out of bowls, and pens crawl across rugs. When Kulagina holds both hands near the motor, the cylinder flies straight up in the air for several meters. A book on the Ripoff rotor (as it is called in Prague), with papers by twelve of the world's leading parapsychologists, is being edited by Ms. Birdbrain.

James Randi, the magician, contends that by using trickery, he can make the motor spin rapidly in either direction. Of course, that does not explain why the motor operates so efficiently for thousands of people who know nothing about conjuring.

ADDENDUM

The foregoing chapter, when it ran in the April 1975 issue of *Scientific American,* was intended as an April Fools' joke. It was so crammed with preposterous ideas and outlandish names that I never dreamed anyone would take it seriously, yet it produced more than a thousand letters from readers who did not recognize the column as a hoax.

The map was designed by correspondent William McGregor (his real name), who gave me permission to print it. Hundreds of readers sent me copies of the map, colored with four colors. Some said they had worked on it for days before they found a way to do it. The four-color map theorem is no longer a conjecture. It was proved in 1976 by Wolfgang Haken and Kenneth Appel, with the aid of a long-running computer program. (*See* "The Solution of the Four-Color-Map Problem," in *Scientific American, October 1977, pp. 108– 121.)* Whether a simple, elegant proof not requiring a computer will ever be found, is still an open question.

When Norman K. Roth published an article, "Map Coloring," in *Mathematics Teacher* (December 1975), many readers informed him that *Scientific American* had published a map disproving the four-color theorem. A letter from Roth in the following May issue pointed out that my column was "an apparently successful April Fools' article."

In 1977 the *Vancouver Sun* reported a British mathematician's claim (it turned out to be invalid) that he had proved the four-color map theorem. On January 17, 1977, the newspaper ran a letter from a lady in Port Moody, which said in part:

> To set the record straight, I would like to bring to your notice the fact that the theorem has already been disproved by William McGregor, a graph theorist . . . in November 1975. He constructed a map of 110 regions that cannot be colored with less than five colors. . . .

The following letter, signed by "Ivan Guffvanoff III," who claimed to be a mathematician at the University of Wisconsin, was a bit frightening to the staff of *Scientific American,* until they realized that it, too, was a joke:

This is to inform you that my lawyer will soon be contacting you for a damage case of $25 million.

In the mathematics section of your April 1975 issue, Martin Gardner wrote that the four-color problem had been solved. I have been working on this problem for 25 years. I had prepared a paper to be submitted to the *American Mathematical Monthly.* The paper was over 300 pages in length. In it I had proved that the answer to the four-color problem was no and that it would take five colors instead of four. Upon reading Gardner's article that someone else would publish the solution before I could, I destroyed my paper. Last week I read in *Time* magazine that Gardner's article was a farce. I did not read Gardner's entire article, only the part on the four-color problem, so I was not aware of the farce. Now that I have destroyed my article, it will not be possible to reproduce all 300 pages, since the work has extended over such a long time. I therefore believe that damages are due me.

I believe that Gardner's article was the most unprofessional article I have ever seen in your's or any other journal. This kind of activity is below the dignity of what I thought your magazine stood for. I am not only suing you but I am cancelling my membership, and I will ask all my friends to cancel theirs.

In Italy the noted mathematician Beniamino Segre published a serious research note (*Rendiconti* 59, 1975, pp. 411–412) in which he reproduced McGregor's map, showing how it could be four-colored. "It is shown the falsehood," his summary reads, "of a presumed counterexample for the four-color conjecture."

Manifold, a journal published by mathematics students at the University of Warwick, ran the following lines in its Autumn 1975 issue. They are to be sung to the tune of "Oh Mr. Porter, what shall I do?"

> "Oh Mr. Gardner,
> What have you done?
> You've started up a rumour
> You should never have begun!
> A four-colour hoax can't
> Be undone so quick . . .
> Oh Mr. Gardner, what
> A bloody silly trick!"

When e is raised to the power of the product of π and the square root of 163, the result is the eighteen-digit number I gave, minus .000,000,000,000,75. . . . John Brillo, to whom I attributed this hoax, is a play on the name of the distinguished number theorist John Brillhart. The reference to Ramanujan's paper is legitimate. In it the Indian mathematician discusses a family of remarkable near-integer numbers to which this one belongs, but of course he knew that none were integral. Indeed, as many readers pointed out, it is not hard to prove that they are transcendental.

The value of the number to thirty-nine significant decimal digits was given by D. H. Lehmer in *Mathematical Tables and Aids to Computation* (vol. 1, January 1943, pp. 30–31). The digit following the run of 9's is 2. See also "What is the Most Amazing Approximate Integer in the Universe?" by I. J. Good, in the *Pi Mu Epsilon Journal* (vol. 5, Fall 1972, pp. 314–315).

The description of Richard Pinkleaf's chess-playing program, MacHic, is a play on the chess program MacHack, written by Richard Greenblatt of the Massachusetts Institute of Technology. The relativity paradox that I hung on Humbert Pringle (a play on the name of Herbert Dingle, a British physicist who maintained that relativity theory is disproved by the famous twin paradox) is well known. It appears as a problem on page 99 of the paperback edition of *Spacetime Physics,* by Edwin F. Taylor and John A. Wheeler (W. H. Freeman and Company, 1966), and the solution is given on page 25 of the answer section. The paradox is discussed at greater length by George Gamow in *Mr. Tompkins in Wonderland* (Macmillan, 1947), W. Rindler in *American Journal of Physics,* (vol. 29, 1961, p. 365 ff.), R. Shaw (*ibid.,* vol. 30, 1962, p. 72 ff.), and P. T. Landsberg in *The Mathematical Gazette,* (vol. 47, 1964, p. 197 ff).

A stationary outside observer will see the meterstick just make it through the hole. If the plate and the stick have thickness, the stick must, of course, be a trifle shorter than the hole to prevent an end from catching. To an observer on the plate the stick will appear Lorentz-contracted, but it will also appear rotated, so that it seems to approach the hole on a slant. The stick's *back* end actually seems to go through the hole before its front end, so that it gets through with the same clearance as before. To an observer on the stick the plate will appear Lorentz-contracted, its hole becoming elliptical, but the plate also appears to be rotated. In this case the hole first goes over the *front* end of the slanted stick, again with the same close fit. "Contractions" and "rotations" are ways of speaking in a Euclidean language. In a four-dimensional, non-Euclidean language of space-time the objects retain their shapes and orientations. Having at one time written a book on relativity, I was abashed to receive more than 100 letters from physicists pointing out the stupid "blunder" I had made.

Those who enjoyed explaining the paradox may wish to consider how to escape from the following variant. Assume that the meterstick is sliding at high speed along the surface of an enormous flat plate of metal toward a hole slightly larger than the stick. We idealize the thought experiment by assuming that there is no friction and that the stick and the plate are extremely thin. When the stick is over the hole, gravity (or some other force) pulls it down and through. For an observer on the stick the sheet slides under it, and the hole is Lorentz-contracted enough to prevent the stick from dropping through. In this case the stick and the plate cannot rotate relative to each other. How does the stick get through? (Please, no letters! I know the answer.)

The Leonardo da Vinci drawing was done by Anthony Ravielli, a graphic artist well known for his superb illustrations in books on sports, science, and mathematics. It was an earlier version of the sketch that suggested to me the idea of a hoax column. Many years ago a friend of Ravielli's had jokingly made a bet with a writer that Leonardo had invented the first valve flush toilet. The friend persuaded Ravielli to do a Leonardo drawing in brown ink on faded paper. It was smuggled into the New York Public Library, stamped with a catalogue file number and placed in an official library envelope. Confronted with this evidence, the writer paid off the bet.

Augusto Macaroni is a play on Augusto Marinoni, a da Vinci specialist at the Catholic University of Milan, and Ramón Paz y Bicuspid is a play on Ramón Paz y Remolar, the man who actually found the two missing da Vinci notebooks. My data on the history of the water closet are accurate, including the reference to Thomas Crapper. The book by Wallace Reyburn *Flushed with Pride: The Story of Thomas Crapper* does exist.

For many years I assumed that Reyburn's book was the funniest plumbing hoax since H. L. Mencken wrote his fake history of the bathtub. I thought this for two reasons: (1) The book implies that the slang words "crap" and "crapper" derived from Mr. Crapper's name, but "crap" and "crapping case" are both listed in *The Slang Dictionary,* published in London in 1873. (2) Reyburn wrote a later book titled *Bust-up: The Uplifting Tale of Otto Titzling and the Development of the Bra.* It turns out, though, that both Thomas Crapper and Otto Titzling were real people, and neither of Reyburn's books is entirely a hoax.

The Ripoff Rotor is a modification of a psychic motor described in Hugo Gernsback's lurid magazine *Science and Invention* (November 1923, p. 651). Prizes were awarded in March 1924 to readers who gave the best explanations of why the cylinder turned. The motion can be caused by any of three forces: slight air currents in the room, convection currents produced by heat from the hand, and currents from breathing. The three forces combine in unpredictable ways. If a person who believes he or she has psychokinetic powers is willing the motor to turn, it may turn in the direction willed, or it may go the other way.

Nandor Fodor, in his *Encyclopedia of Psychic Science* (Citadel, 1966), under the heading "Fluid Motor," credits the paper-cylinder device to one Count de Tromelin, but he doesn't say who the Count was or when he invented the motor.

I had no individual in mind when I mentioned Ms. Henrietta Birdbrain, but there *is* an *East West* newspaper in Boston, and the reader who bothers to check the issue cited, will find a sober report by Stanley Krippner on a psychic motor that was demonstrated to him in Prague by Robert Pavalita. (On Pavalita, see Chapter 28 in *Psychic Discoveries behind the Iron Curtain* (Prentice-Hall, 1970), by Sheila Ostrander and Lynn Schroeder.)

Many readers, in the spirit of the hoax, sent hilarious explanations for the Ripoff Rotor. Mark J. Hagmann found that the rate of the relative rotation of the rotor and the room was a function of the contents of the liquor bottle he used to support the needle. The rotation increased as the level of the liquid went down.

In concluding this addendum, let me pass along some sage advice supplied by my brother Jim. Every responsible science writer should constantly keep in mind the following four words: "Accuracy above all."

ELEVEN

:
:
:

The Császár Polyhedron

Donald W. Crowe, a mathematician at the University of Wisconsin, discovered a surprising correspondence between the skeletons of *n*-dimensional cubes and the solution of a classic puzzle called the Tower of Hanoi. I described his discovery in a column reprinted in *The Scientific American Book of Mathematical Puzzles & Diversions* (Simon & Schuster, 1959). Professor Crowe has now done it again. In studying the skeletal structure of a bizarre solid known as the Császár polyhedron, he has found some remarkable isomorphisms that involve the seven-color map on a torus, the smallest "finite projective plane," the solution of an old puzzle about triplets of seven girls, the solution of a bridge-tournament problem about eight teams, and the construction of a new kind of magic square known as a Room square.

The polyhedron that leads into so many apparently unrelated recreations is enormously interesting in itself. It is the only known polyhedron, apart from the tetrahedron, that has no diagonals. (A diagonal is a line joining any two vertexes not connected by an edge.) Consider, for example, the tetrahedron. It has four vertexes, six edges, four faces, and no diagonals. An edge joins every pair of corners.

Skeletons of polyhedrons are isomorphic with graphs, that is, with sets of points (vertexes) joined by lines (edges). If an edge connects every pair of points in a set of *n* points, it is called a complete graph for *n* points. Any polyhedron without diagonals clearly must have a skeleton that is a complete graph. Since no polyhedron can have fewer than four corners, the complete graph for four points is the simplest such graph that corresponds to a polyhedral skeleton.

We must now be careful to make precise definitions. A simple polyhedron is one that is topologically equivalent to a sphere and whose faces are all simple polygons: polygons topologically equivalent to a disk. (If you think of a simple polyhedron as having elastic faces, the polyhedron can be inflated like a balloon to form a sphere.) This rules out such nonsimple structures as two polyhedrons joined by an edge or a corner, stellated polyhedrons with intersecting faces, solids with tunnels or interior holes, and so on. Imagine a polyhedron with a surface like that of a cube with a smaller cube on the center of one face. The polyhedron can be inflated to a sphere, but is it simple? It is not, because one face is a ring.

The tetrahedron is the only simple polyhedron with no diagonals. An interesting question now arises. Let us define a toroid as a polyhedron whose faces are all simple polygons, but the solid itself is topologically equivalent to a sphere with one tunnel or more going all the way through it. Is it possible to construct a toroid with no diagonals?

The answer to the question was not known until the late 1940s, when a Hungarian topologist, Ákos Császár, succeeded in constructing such a polyhedron. Its skeleton is the complete graph for seven points, shown in symmetrical form at the left of Figure 61. The graph is isomorphic with the skeleton of a six-dimensional simplex, the 6-space analogue of the tetrahedron. Since an edge joins every pair of points, the polyhedron has twenty-one edges and fourteen triangular faces.

It is not difficult to make a paper model of the Császár polyhedron. Copy the two patterns in Figure 62 on paper of good quality (or thin cardboard) and cut the copies out. Color the seven shaded triangles on both sides. Crease the paper to make "mountain folds" along each broken line, and "valley folds" along each solid line.

1. With the pattern for the base, fold the two largest triangles to the center and tape the *A* edges to each other. Turn the paper over. Fold the two smaller triangles to the center and tape the *B* edges together to obtain a completed base.

2. The six-faced conical top is formed by taping the *C* edges together. Place it on the base as shown in the drawing of the completed model. It will fit

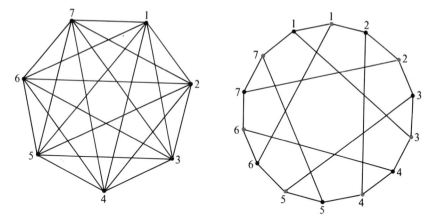

Figure 61 Skeleton of the Császár polyhedron (left) and its dual, the seven-color torus map (right)

in two ways. Choose the fit that joins white to shaded triangles, then tape each of its six edges to the corresponding six edges of the base.

It is not yet known if there is another toroid without diagonals. By applying elementary Diophantine analysis (finding integral solutions of equations), however, Crowe has shown that if there is another, it will have at least twelve vertexes and six tunnels. His proof is as follows:

For simple polyhedrons there is a famous formula, discovered by Leonhard Euler, that relates the vertexes (v), edges (e) and faces (f). It is

$$v - e + f = 2$$

The formula is easily proved, and it takes only a bit of extra work to modify it for toroids. Letting h stand for holes, the formula is

$$v - e + f = 2 - 2h$$

If a toroid is without diagonals, its skeleton is a complete graph, and for any complete graph, edges and vertexes are related by the formula

$$e = \tfrac{1}{2}v(v - 1)$$

PATTERN FOR TOP

PATTERN FOR BASE

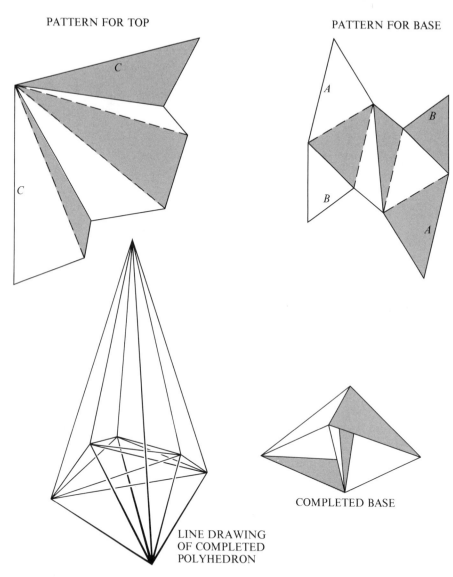

COMPLETED BASE

LINE DRAWING
OF COMPLETED
POLYHEDRON

Figure 62 Patterns for making a model of the Császár polyhedron

The faces and edges of a toroid without diagonals (all its faces must be triangles, otherwise a face would have a diagonal on it) are related by the formula

$$f = 2e/3$$

Substituting the values for e and f in the formula $v - e + f = 2 - 2h$ and simplifying, we get

$$12h = v^2 - 7v + 12$$

Factoring the right side gives h a value of

$$h = \frac{(v - 3)(v - 4)}{12}$$

The values of v and h must be integers, and we also know that v must be greater than 3. If v is 4, h has a value of zero. These values fit the tetrahedron. If v is 5 or 6, h is not integral, proving that no toroid without diagonals can have five or six corners. When v is 7, h is 1. This corresponds to the Császár polyhedron. The next solution in integers is $v = 12$, $h = 6$. Whether a toroid can be constructed for these values remains unknown. Nor is it known if toroids exist for the next two solutions: $v = 15$, $h = 11$, and $v = 16$, $h = 13$. From here on, Crowe points out, the number of tunnels exceeds the number of vertexes, so that we can probably rule out all higher toroids.

The Császár polyhedron has one tunnel, which means that if our model's surface consisted of rubber rather than paper, we could blow it up to the shape of an inner tube. Its twenty-one edges would then form a complete graph for seven points on the surface of the torus. None of these edges would intersect one another, proving that the complete graph of seven points has a toroidal crossing number of zero. (On crossing numbers, see Chapter 11 of my *Knotted Doughnuts and Other Mathematical Amusements,* W. H. Freeman and Company, 1986).

Suppose now that on this torus we change the complete graph to its dual. That is done by putting a point inside each of the fourteen triangular faces, then drawing an edge from each point to the points inside its three neighboring faces. Since each new edge crosses one old edge, the number of new edges remains twenty-one. The number of faces and the number of vertexes, however, are switched. The new graph is shown in a symmetrical planar form at the right in Figure 61. Crowe has made the fourteen vertexes alternately gray and black spots and has numbered them, as shown, for reasons that will shortly be clear. Its "faces" are hard to see, but you can trace out seven regions, each region surrounded by six edges.

This graph can also be drawn on a torus without any intersecting edges. When that is done, the result turns out to be the familiar seven-color toroidal map (see Figure 63). Note that every two of the seven hexagonal regions share

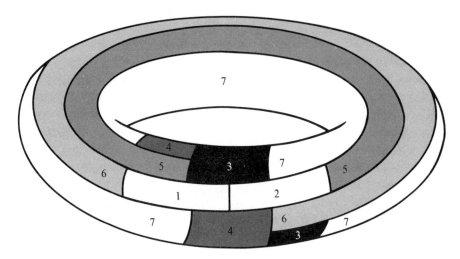

Figure 63 A seven-color map on the torus (regions 3, 4, and 7 wrap around)

a common edge. This means that if the map is colored so that no two contiguous regions have the same color, seven colors are required. On a plane, no more than four regions can be mutually contiguous, but on a torus the maximum is seven.

As an exercise, the reader may like to put a dot inside each region on the torus and see if the map can be converted to its dual by the same procedure described above. Simply draw an edge to connect each pair of dots. Each edge must cross just one boundary segment, and there must be no intersecting of the new edges with each other. Some lines must, of course, wrap around the torus. If you succeed, the new graph will be isomorphic with the complete graph of seven points and also with the skeleton of the Császár polyhedron.

Here is how Crowe used the dual graph of the skeleton of the Császár polyhedron to solve the following puzzle. Seven girls live in a house. Each day a triplet of girls is allowed to leave the house for a visit to town. How can the triplets be chosen so that at the end of seven days every pair of girls will have been in exactly one of the seven triplets?

The graph at the right in Figure 61 provides two solutions. For one solution, note each black vertex and write down the (unordered) triplet of numbers on the three gray vertexes that are adjacent (connected by an edge) to it. For the other solution write down the numbers on the black vertexes adjacent to each gray vertex. The two sets of triplets are

Black vertex	Gray vertex
124	126
235	237
346	341
457	452
561	563
672	674
713	715

Each set is called a "Steiner triple system" of order 7, or a "finite projective plane" of order 2. Steiner systems and projective planes are topics of great importance in modern combinatorial theory, but we can (regrettably) mention them only in passing.

Because the seven-color graph is the dual of the Császár skeleton, with each of its fourteen vertexes corresponding to one of the fourteen faces of the model of the Császár polyhedron (black spots to white faces, gray spots to shaded faces), we can just as easily extract the two solutions from the model. On the model, no face shares a border with a face of the same color. If the faces are numbered to correspond to the numbers on the vertexes of the graph, the triplet of numbers on the three white faces adjacent to each shaded face gives one solution, and the triplet of numbers on the three shaded faces adjacent to each white face gives the second solution.

Another way to find the same two sets of triplets on the model is to number the vertexes of the model, any way you like, from 1 through 7. The numbers at the three corners of each shaded face give one set of triplets, and those at the three corners of each white face give the other set. The two solutions will be equivalent to the two sets of triplets listed above, although the numbers may not match. The numbers are no more than arbitrary symbols on a symmetrical graph. To see the equivalence, it may be necessary to permute them in some way, such as changing all 1's to 5's, all 5's to 3's, and so on.

The two solutions obtained by any of these methods are also identical in the sense that one can be changed to the other by permuting the elements; in other words, there is only one basic solution. The next higher Steiner triple system is of order 9. It too has a unique basic solution. Nine girls go out in daily triplets for twelve days, each pair appearing in just one triplet. Can the reader find the solution?

The two variants we obtained for the order-7 solution are, as Crowe recognized, related in a curious way. No triplet appears in both sets, and if two pairs of girls in one set appear with the same third girl (for example, in the first set 1,2 and 3,6 each appear with 4), the same pair appears with different girls (5,6) in the second set. When both of these properties hold, the two solutions are called orthogonal.

A Steiner triple system of order n is possible only when n is equal to 1 or 3 (modulo 6). Every orthogonal pair of such systems of order n, Crowe goes on to explain, provides a solution to the following bridge problem for $n + 1$ teams. Suppose there are eight teams of card players and seven tables. Each team must play exactly once with each of the other teams and also exactly once at each table.

This is how Crowe tells us to construct the tournament. First draw a square matrix seven cells by seven cells. Consider the pair 1,2. In the first set of triplets, it is associated with 4 and in the second set with 6, so that we put 1,2 in the cell at the intersection of the fourth column and the sixth row (*see* Figure 64). Consider another pair, 1,3. It is with 7 in the first set and with 4 in the other, so that 1,3 goes in the seventh column and the fourth row. Follow this procedure for all pairs of numbers. The final step is to combine 8 with 1, 2, 3, 4, 5, 6, 7 along a diagonal from the cell at the upper left to the cell at the lower right. Each column indicates a table, and each row indicates a round of simultaneous play at four of the seven tables. All conditions of the desired tournament are now met.

The matrix is called a Room square of order 8. Such a square is an arrangement of an even number of objects, $n + 1$, in a square array of side n. Each cell is either empty or holds exactly two different objects. In addition, each object appears exactly once in every row and column, and each (unordered) pair of objects must occur in exactly one cell.

The smallest Room square is trivial. It is of order 2 and consists of one cell that contains 1,2. No Room squares are possible for four or six objects, so that the order-8 square is the smallest nontrivial Room square.

For years, I assumed that such squares were called Room squares because they concern objects placed in "rooms," but it turns out that they are named for Thomas G. Room, a mathematician who defined them in a brief note in 1955. Combinatorialists have been investigating them ever since. Later it was discovered that Room squares had been in use before 1900 in bridge tournaments, but mathematicians seem not to have been interested until Room wrote his note.

There is still more! H. S. M. Coxeter, in editing the twelfth revised edition of W. W. Rouse Ball's classic *Mathematical Recreations and Essays* (University of Toronto Press, 1974), explains how the Steiner triple system of order 7 can be used for constructing an "anallagmatic pavement" of order 8. Consider the order-8 chessboard. If we place any two rows alongside each other, either every cell in one row will match the color of its neighbor in the other, or every

TABLES

	1	2	3	4	5	6	7
1	1,8			5,7		3,4	2,6
2	3,7	2,8			6,1		4,5
3	5,6	4,1	3,8			7,2	
4		6,7	5,2	4,8			1,3
1	2,4		7,1	6,3	5,8		
6		3,5		1,2	7,4	6,8	
7			4,6		2,3	1,5	7,8

Figure 64   The smallest nontrivial Room square

cell will not match its neighbor's color. We want to color the sixty-four cells with two colors so that the following property holds: If any two rows are brought together, half of the paired cells will match and half will not, and the same will be true of any two columns.

Such squares are also known as Hadamard matrixes, after the French mathematician Jacques Hadamard, who studied them in the 1890s. Apart from the trivial case of order 2, no Hadamard matrix is possible unless the order is a multiple of 4. It is not yet known if matrixes for all such orders exist.

Figure 65 shows how our first Steiner triple system provides a Hadamard matrix for the chessboard. The problem involves eight girls and eight days.

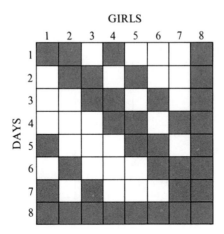

Figure 65   A Hadamard chessboard

Number the rows and columns as shown. The eighth girl is an older girl who chaperones the triplets daily, and on the eighth day all eight girls go into town. For each day (indicated by a row), color the cells that indicate the three girls (plus the chaperone) who walk to town. The result: a Hadamard matrix!

There is a simple technique for generating Hadamard matrixes for all orders that are powers of two (*see* Figure 66). The order-2 pattern is placed in three corners of the order-4 square, and its negative (colors reversed) goes into the lower right corner. The same procedure with the order-4 pattern generates the order 8, and so on for the higher powers.

More generally, given two Hadamard matrixes of orders *m* and *n,* a matrix of order *mn* can be created simply by replacing each colored cell of *m* by the entire pattern of *n* and each uncolored cell of *m* by the negative of *n*. The large matrix is called the tensor product of the two smaller ones. It does not matter whether *m* is larger, smaller, or equal to *n*. If, for example, you have a Hadamard matrix of order 12, the tensor product of two such matrixes is a Hadamard matrix of order 144.

Hadamard matrixes are more than playthings. They are used for the construction of valuable error-correcting binary codes. For descriptions of such applications, the reader is referred to the new edition of Ball's book (so heavily revised as to be almost a new work) and to the references it cites. When the Mariner spacecraft of 1969 sent back pictures of Mars, Coxeter tells us, they were sent in an error-correcting code based on the order-8 Hadamard matrix.

In closing, let us return to the Császár polyhedron and pose an intriguing problem. The Császár toroid cannot be constructed if all its faces are equilateral triangles. Suppose a one-hole toroid is made entirely of congruent triangular faces, all equilateral. What is the minimum number of faces it can have? Bonnie M. Stewart, a mathematician at Michigan State University, considers this problem on page 48 of his book *Adventures among the Toroids* (1970).

Stewart gives the construction of such a torus with fifty-four faces, twenty-seven vertexes, and eighty-one edges. One of his students, Kurt Schmucker, found a one-hole toroid with forty-eight equilateral triangular faces. Whether that is the minimum, however, is another tantalizing, unanswered toroidal question.

ANSWERS

The unique solution for the Steiner triple system of order 9 is obtained from the following matrix:

$$\begin{matrix} a & b & c \\ d & e & f \\ g & h & i \end{matrix}$$

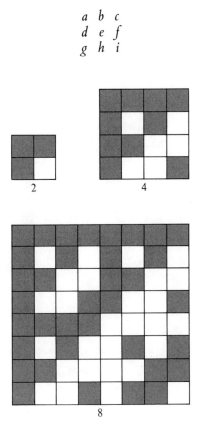

Figure 66 Hadamard matrixes for powers of two

The rows (*abc, def, ghi*) provide three triplets. The columns (*adg, beh, cfi*) provide three more. The main diagonals (*aei, ceg*) and the broken diagonals (*bfg, cdh, afh, bdi*) complete the list of twelve triplets.

The orders of Steiner triple systems are numbers that have a remainder of 1 or 3 when they are divided by 6. The next higher system, order 13, has two basic solutions. Order 15 is known to have eighty solutions.

A model of Kurt Schmucker's one-hole toroid, with forty-eight congruent equilateral triangle faces, is easy to make. It consists of a ring of eight regular octahedrons joined by their faces (*see* Figure 67). Schmucker found that rings could be made by joining eight replicas of each of the Platonic solids except the tetrahedron. No matter how many tetrahedrons are joined by their faces, no ring is possible even when the solids are allowed to intersect one another. A proof is given by J. H. Mason in his paper "Can Regular Tetrahedra Be Glued Together Face to Face to Form a Ring?" (*The Mathematical Gazette* 56, October 1972, pp. 194–197).

Figure 67 Ring of eight octahedrons

ADDENDUM

When I wrote briefly about Hadamard matrixes, I had not realized that their application to telemetry codes was discovered by Golomb and that the use of such codes in several Mariner probes of Mars was the result of Golomb's affiliation with the Jet Propulsion Laboratory of the California Institute of Technology (*see* Bibliography for Golomb's co-authored 1963 paper). A Hadamard matrix larger than 2 by 2 must have a side that is a multiple of 4, but it is not yet known if matrixes exist for all such orders. Golomb tells me that a 268 Hadamard matrix was reported by Kanue Sawade in *Graphs and Combinatorics* (vol. 1, 1985, pp. 185–187). The next unknown case, now the smallest, is order 428.

Hadamard matrixes have also found extensive application in the processing of pictorial information. The "Hadamard transform" (analogous to the "fast Fourier transform") actually produces a mathematical hologram of the original image. Golomb has been a pioneer in recognizing that classic combinatorial design problems often provide optimum solutions to data-processing problems and that it pays to look for engineering problems to which these designs are the solution.

B I B L I O G R A P H Y

"A Polyhedron without Diagonals." Ákos Császár, in *Acta Scientiarum Mathematicarum* 13, 1949–1950, pp. 140–142.

Adventures Among the Toroids. B. M. Stewart. Published by the author, 1952. 2d rev. ed., 1984.

"The Search for Hadamard Matrices." Solomon W. Golomb and Leonard D. Baumert, in *American Mathematical Monthly* 70, January 1963, pp. 12–17.

"Euler's Formula for Polyhedra and Related Topics." Donald W. Crowe, in *Excursions in Mathematics*, Anatole Beck et al., eds. Worth, 1969.

"Steiner Triple Systems, Heawood's Torus Coloring, Császár's Polyhedron, Room Designs, and Bridge Tournaments." Donald W. Crowe, in *Delta* 3, Spring 1972, pp. 27–32.

Room Squares

"A New Type of Magic Square." Thomas G. Room, in *The Mathematical Gazette* 39, 1955, p. 307.

"On Furnishing Room Squares." R. C. Mullin and E. Nemeth, in *Journal of Combinatorial Theory* 7, November 1969, pp. 266–272.

"On Room Squares of Order $6m + 2$." C. D. O'Shaughnessy, *ibid,* vol. 13, series A. November 1972, pp. 306–314.

"Solution of the Room Square Existence Problem." W. D. Wallis, *ibid,* vol. 17, series A Novemer 1974, pp. 379–383.

"The Existence of Room Squares." R. C. Mullin and W. D. Wallis, in *Aequationes Mathematicae* 13, 1975, pp. 1–7.

TWELVE

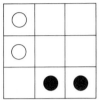

.

.

.

Dodgem and Other Simple Games

Consider a two-person game with the following characteristics: (1) It is a game of perfect information; that is, both players have complete knowledge of the game's structure after every move. (2) The players move alternately. (3) Decisions are not made by chance. (4) The game ends after a finite number of moves with a win by one player. (No draw is possible.)

It is not hard to see that there must be a winning strategy for either the first player or the second. If the first player (henceforth called *A*) does not have a winning strategy, he must lose. This means that the second player, *B*, has a winning strategy. Does the argument apply if we rescind the requirement that the game end in a finite number of moves?

Curiously, it all depends on whether or not one accepts the "axiom of choice." This notorious axiom says that from any collection (finite or infinite) of nonempty sets, with no elements in common, you can form a new set by taking one element from each set. In the 1930s, Stefan Banach, Stanislaw Mazur, and Stanislaw Ulam discovered a type of infinite game in which neither *A* nor *B* has a winning strategy if the axiom of choice is accepted. Someone argued that this proves the Unitarian dogma that there is "at the most" one God, because if two gods could play such a game, neither could know a winning strategy and therefore neither could be called omniscient!

That, however, is by the way. Here we shall examine some new two-person nonchance games for which the rules are extremely simple and for which a winning strategy is either known or capable of being known. All but one of the games are played with counters on boards that are easily drawn on cardboard. Two differently colored sets of counters, such as go stones or small poker chips, will be useful for any reader who wants to play or to analyze the games.

An example of an almost trivial game of the nim type, but one with a strategy that is not immediately apparent, is played on the star pattern shown in Figure 68. Put a counter on each of the star's nine points. A and B take turns removing either one counter or any two counters joined by a straight line segment. The player who takes the last counter wins.

B can always win at star nim by a strategy based on the board's symmetry. Imagine that the black lines are strings. The pattern can be opened up to a circle that is topologically equivalent to the star. If A takes one counter from this circle, B takes the two counters that are directly opposite. If A takes two counters, B takes one counter that is directly opposite. In each case two sets of three counters are left. Now, whatever A takes from one set, B takes the corresponding counter or counters from the other set. Obviously B will get the last counter. If the reader plays a few games on the circle, translating each move to its equivalent on the star, he or she will soon see how to use the star's symmetry for playing the strategy.

In the late 1960s, G. W. Lewthwaite, of Thurso, Scotland, invented a delightful game with an artfully concealed "pairing strategy" that gives the second player a sure win. On a 5-by-5 square matrix place thirteen black

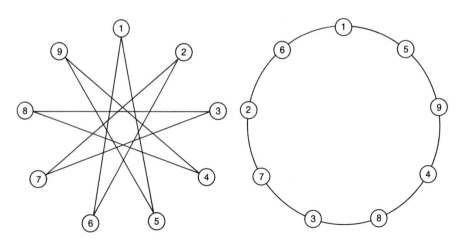

Figure 68 Star nim (left) and its winning strategy (right)

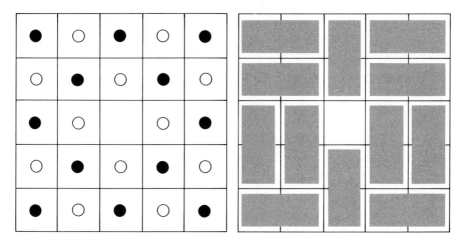

Figure 69 G. W. Lewthwaite's counter game (left) and a pairing strategy for Lewthwaite's game (right)

counters and twelve white counters in alternating checkerboard fashion. Any one of the black counters, say the one in the center, is removed (*see* Figure 69, left).

Player *A* controls the white counters and *B* the black. They take turns moving one of their counters orthogonally to the vacant square until a player loses by being unable to move. If the board is colored like a checkerboard, it is obvious that on each move, a counter goes to a square of different color and that no counter can be moved twice. The game, therefore, cannot go beyond twelve moves for each player. It may end before then, however, in a win for either player unless *B* plays rationally.

B's strategy is to imagine that the matrix, except for the initially vacant cell, is covered with twelve nonoverlapping dominoes. It does not matter how they are placed. Figure 69 (right) shows a sample pattern. Whenever *A* moves, *B* simply moves his counter that is on the domino *A* has just vacated. Since this ensures that *B* always has a move to follow a move by *A, B* is sure to win in twelve or fewer moves.

The game can be played not only with counters but also with square tiles or cubes that slide within a matrix surrounded by a rim. Suppose the rules are amended to allow either player at any time to move any number of adjacent counters (1 through 4) in a row or column provided that the two end counters are of his or her color. This is a splendid example of how an apparently trivial alteration of a rule can enormously complicate a game's analysis. Lewthwaite was unable to find a winning strategy for either player in this variant of his game.

Games based on the sliding of unit squares within a square matrix offer a plethora of unexplored possibilities. Lewthwaite proposes an attractive game that he calls meander. It uses twenty-four identical tiles placed in a 5-by-5 tray to form the pattern shown in Figure 70 (left). The players take turns sliding a single counter or a straight row or column of two, three, or four counters. The play continues until a player wins by creating a pattern in which at least three tiles form a continuous line or path that joins two edges (opposite or adjacent) of the tray. Figure 70 (right) shows a winning pattern, with the winning line indicated by the two arrows. The game is probably too complex for solving without a computer program, and perhaps too complex for solving even with one.

In 1972, when Colin Vout was a mathematics student at the University of Cambridge, he invented an intriguing counter game that he calls dodgem because it is so often necessary for a piece to dodge around enemy pieces. It is playable on a checkerboard of any size. Even the game on a 3-by-3 board is complicated enough to be interesting.

Two black counters and two white ones are initially placed as shown in Figure 71 (top). Black sits on the south side of the board and White sits on the west. The players alternately move a counter one cell forward or to their left or right, unless it is blocked by another counter of either color or by an edge of

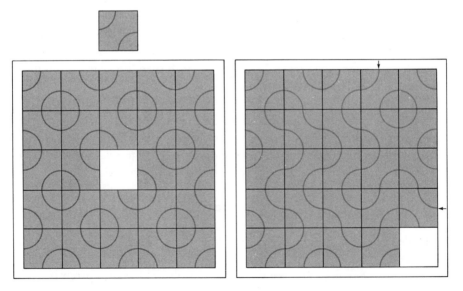

Figure 70 Meander, with example of pattern on a tile at top (left) and a possible winning pattern in meander (right)

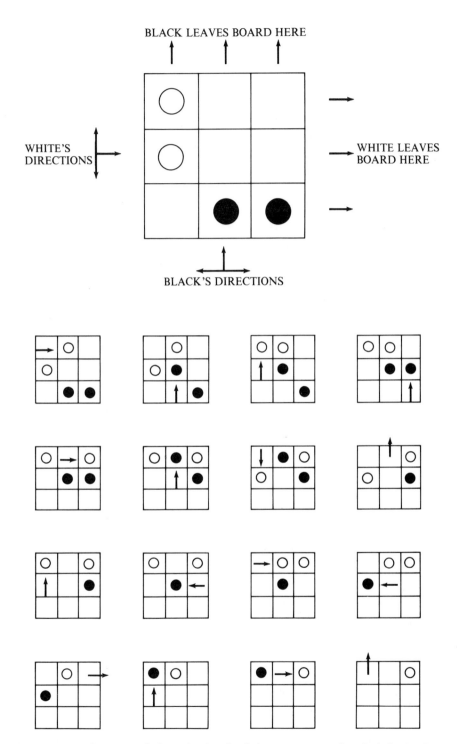

Figure 71 Colin Vout's dodgem (top) and a dodgem game won by Black (bottom)

the board. Each player's goal is to move all his pieces off the far side of the board. In other words, Black moves orthogonally north, west, or east and attempts to move both of his pieces off the north side of the board. White moves east, north or south and tries to move his pieces off the east side of the board.

There are no captures. A player must always leave his opponent a legal move or else forfeit the game. The first to get all his pieces off the board wins. The bottom of Figure 71 shows a typical game won by Black.

Vout assures me that the first player has the win on the order-3 board, but as far as I know, no games on higher-order boards have yet been solved. On a board of side n, each player has $n - 1$ pieces placed on the west and south edges, with the southwest corner cell vacant. Played with seven checkers or pawns of one color and seven of another color on the standard order-8 checkerboard or chessboard, it is a most enjoyable game.

Piet Hein's now classic game of hex (see Chapter 8 of my *Scientific American Book of Mathematical Puzzles & Diversions*, Simon & Schuster, 1959) remains unsolved, except for small boards. For readers unfamiliar with the game, it is played on an n-by-n rhombus of hexagons such as the order-4 board shown in Figure 72. White opens by placing a white counter on a cell. Black follows with a black counter. They take turns placing counters on vacant cells (there are no moves or captures) until a player wins by forming a chain of adjacent counters that joins his side of the board to the opposite side, White by joining the north and south edges, Black by joining the east and west edges.

It is easy to see that no draw is possible. There is a famous proof by John F. Nash (who independently invented hex) that on a rhombus of any size, the first player has a winning strategy, although the proof gives no hint of what the strategy is.

Suppose White allows Black to tell him where he must make his first move. Can White still always win if he plays rationally? This modified version of hex has been called Beck's hex after Anatole Beck, who both proposed and solved it. Writing on hex in Chapter 5 of *Excursions into Mathematics*, Beck shows that Black can always win if he tells White to open by taking an acute corner cell. In other words, such an opening is a sure loss for White, although Beck's proof does not provide Black's winning strategy. However, as a footnote comments, it "wrecks Beck's hex."

What about *misère*, or reverse, hex, known as rex, in which the first player to join his sides loses? As is so often the case in two-person games, the reverse game proves to be much harder to crack. No general strategy is known, although Robert O. Winder, in unpublished arguments, has shown the existence of a first-player winning strategy in rex of even order and a second-player winning strategy for all odd orders. More recently Ronald J. Evans has

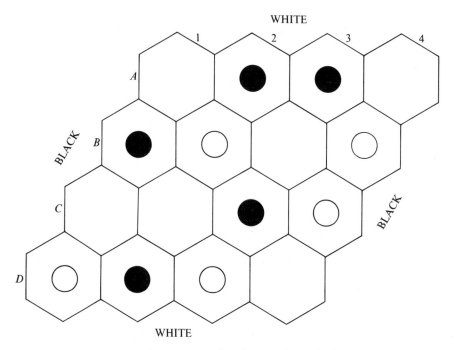

Figure 72   Rex, a reverse hex game, with White to play and win .

carried Winder's arguments a step further by showing that on even-order boards there is a winning strategy if White opens in the acute corner.

Rex on the order-2 rhombus is trivial, and it is not difficult to analyze exhaustively on the order-3 rhombus. Play on the order-4 rhombus is so complicated, however, that even though it is known that an acute-corner opening initiates a win, the strategy itself remains unformulated. The position shown in Figure 72 is an order-4 rex problem composed by Evans. Can the reader determine White's only correct move?

Here is an even simpler game for which no general strategy is known. It is played on a 1-by-n board (a single row of n squares) with counters that are all alike. A and B take turns placing a counter until one player wins by getting three counters adjacent. Could anything be simpler? A can always win when n is odd by first taking the center cell, then playing symmetrically opposite the opponent thereafter. For even n, however, things are not so simple. On most even rows, A seems to have the win, but not necessarily, and the exceptions follow no known rule. Take $n = 6$, for example. The reader may enjoy working it out to see who has the win.

John Horton Conway has pointed out that this game is equivalent to a game I called 1-by-n cram in a column reprinted as Chapter 19 of my *Knotted*

Doughnuts and Other Mathematical Entertainments, (W. H. Freeman and Company, 1986) except that it is played with trominoes instead of dominoes. It is easy to see the isomorphism. In playing the game as described above, it is obviously disastrous to place a counter either next to another counter or one cell from it, since either move gives the opponent an instant win. Hence we might as well prohibit both moves. An easy way to do it is to require that each play consist of a triplet of adjacent counters, which is the same as placing a tromino on the field. (The middle of the triplet corresponds to placing a single counter, and the ends of the triplet enforce the two new rules.) The winner is the player who places a tromino last. (To complete the equivalence, we must allow the placing of a tromino at either end of the field so that it extends one cell beyond the end.) Of course, the game can also be played by forming a row of *n* counters and by removing them by alternate moves of taking three adjacent counters.

This triplet version of cram is considered in a classic paper by Richard K. Guy and Cedric A. B. Smith, "The G-Values of Various Games. Because it is coded as game .007, it has been called the James Bond game. Elwyn Berlekamp has computer-analyzed the game to very high even *n* without finding any periodicity in the Grundy numbers, which means that no one is even close yet to a general rule. The *misère* version of 1-by-*n* tromino cram, regardless of the parity of *n,* is also unsolved.

Ulam has proposed extending the counter form of tromino cram to a square matrix. The players take turns placing single counters until one player wins by getting three in a row orthogonally or diagonally. As before, odd-order fields are trivial because the first player wins by taking the center, then playing symmetrically until the opponent offers a win. On even-order boards the order 4 is trivial, but no one yet knows who has the win on orders 6 or 8. Figure 73, supplied by Ulam in a letter, shows a position on the order-6 board for which the next player must lose.

Here again we can play an equivalent game by alternately placing polyominoes, in this case squares of nine cells, but it is not very convenient because in addition to allowing the pieces to extend into a unit border around the field, we must also allow them to overlap one another by just two cells (corner and side). No one has even begun to find a general strategy for the game in standard or reverse form.

ANSWERS

The answers to the chapter's two problems are that the second player has the win on 1-by-6 tromino cram, and White wins the 4-by-4 rex (reverse hex)

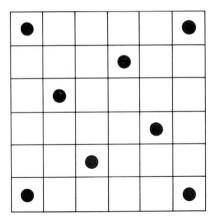

Figure 73   Stanislaw Ulam's triplet game

game by taking the cell at the intersection of row *C* and diagonal 1. Although no winning strategy for the first player of order-4 rex is known, a winning second-player pairing strategy for the order-5 game has been found by David L. Silverman. All higher orders remain unsolved.

G. W. Lewthwaite's game generalizes in an obvious way to rectangular boards of any size and shape. If the rectangle has an odd number of cells, the second player wins; if it has an even number of cells, the first player wins. (In the latter case the domino covering strategy includes the initially vacant cell.)

Karl Fulves has suggested that instead of visualizing a domino pattern, you play the game with counters secretly marked so that you can place them on the board in orientations that group them in pairs. For example, a small pinhole on the rim of a pawn or a checker would enable you to orient the pieces so that the pinholes of each pair are adjacent. You could then play the domino strategy without having to remember a domino pattern. If coins are used, designate a spot on the rim of each side of each coin, say the N in the ONE on a penny, as your mark and orient the coins accordingly.

ADDENDUM

David Fremlin and Dennis Rebertus independently wrote computer programs that verified the first player's win in order-3 dodgem. White wins only by first moving his piece at the corner. A full analysis of the order-3 game is in *Winning Ways.* Order-4 dodgem is still unsolved.

	3	4	5	6	7	8	9	10	11	12
3	2,2	*	2,3	*	2,4	2,5	2,5	*	2,6	3,3
4		3,3	*	3,4	4,1	*	*	3,8		
5			3,3	4,1	3,4	3,5	3,5			
6				4,4						

Figure 74 John Beidler's results for Stanislaw Ulam's triplet game

John Beidler, who heads the computer science department at the University of Scranton, found by computer that Stanislaw Ulam's triplet game in standard play on a 6-by-6 field is a win for the first player only if his first move is on one of the four central cells. Beidler generalized the game to rectangular boards and obtained the results shown in Figure 74. The numbers give winning moves by row and column for the first player. The asterisks indicate a win for the second player. If the game is played in reverse form, Beidler found that the second player has the win on 3-by-3, 3-by-4, 3-by-5, and 4-by-4 boards. For tromino cram, played in reverse on a 1-by-n field, with n less than 19, Beidler proved first-player wins for $n = 4, 5, 7, 8, 11, 14, 15, 17,$ and 18, and second-player wins for all other values.

B I B L I O G R A P H Y

"The G-Values of Various Games." R. K. Guy and C. A. B. Smith, in *Proceedings of the Cambridge Philosophical Society* 52, July 1956, pp. 514–526.

"The Game of Hex." Anatole Beck, in *Excursions into Mathematics,* Beck et al., eds. Worth, 1969.

"A Winning Opening in Reverse Hex." R. J. Evans, in *Journal of Recreational Mathematics* 7, Summer 1974, pp. 189–192.

"Reverse Cram with Block Sizes Not Exceeding 13." R. J. Evans, in *Delta* 6, 1976, pp. 57–76.

"Dodgem." *Winning Ways.* E. R. Berlekamp, J. H. Conway, and R. K. Guy, pp. 685–688. Academic Press, 1982.

THIRTEEN

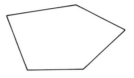

.
.
.

Tiling with Convex Polygons

"Many of the brightly coloured,
tile-covered walls and floors of the
Alhambra in Spain show us that the
Moors were masters in the art of
filling a plane with similar
interlocking figures, bordering each
other without gaps. What a pity that
their religion forbade them to make
images!"

— M. C. Escher

Imagine that you have an infinite supply of jigsaw puzzle pieces, all identical. If it is possible to fit them together without gaps or overlaps to cover the entire plane, the piece is said to tile the plane, and the resulting pattern is called a tessellation. From the most ancient times such tessellations have been used throughout the world for floor and wall coverings and as patterns for furniture, rugs, tapestries, quilts, clothing, and other objects. The Dutch artist M.

C. Escher amused himself by tessellating the plane with intricate shapes that resemble birds, fish, animals, and other living creatures (*see* Figure 75).

A tile that tessellates obviously can have an infinite variety of shapes, but by imposing severe restrictions on the shape, the task of classifying and enumerating tessellations is reduced to something manageable. Geometers have been particularly interested in polygonal tiles, of which even the simplest present formidable problems. In this chapter we are concerned only with the task of finding all convex polygons that tile the plane. It is a task that was not completed until 1967, when Richard Brandon Kershner, assistant director of the Applied Physics Laboratory of Johns Hopkins University, found three pentagonal tilers that had been missed by all predecessors who had worked on the problem.

Let us begin by asking how many of the regular polygons tile the plane. As the ancient Greeks knew and proved, there are just three: the equilateral triangle, the square, and the regular hexagon. The hexagonal tiling, so familiar to bees and users of bathrooms, is a fixed pattern (*see* Figure 76). The patterns formed by equilateral triangles or by squares can be infinitely varied by sliding rows of triangles or squares along lattice lines.

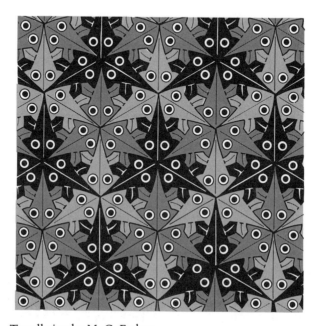

Figure 75 Tessellation by M. C. Escher

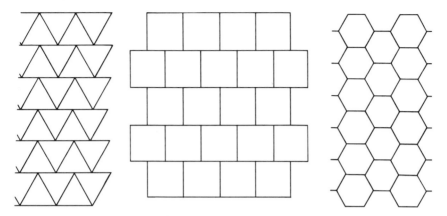

Figure 76   The three regular polygons that tile the plane

If we remove the restriction that a convex polygon must be regular, the tiling problem grows in interest. It has been proved that no convex polygon of more than six sides can tile the plane. Thus we need to investigate only polygons of three, four, five, and six sides.

The triangle is easy. Any triangle tiles the plane. Simply fit two identical triangles together, with the corresponding edges coinciding as shown in Figure 77, and you create a parallelogram. Replicas of any parallelogram obviously will go side by side to make an endless strip with parallel sides, and the strips, in turn, go side by side to fill the plane.

The quadrilateral is almost as easy, although much more surprising. Any quadrilateral tiles the plane! As before, take a pair of identical quadrilaterals, one inverted with respect to the other, join the corresponding edges and you create a hexagon (*see* Figure 78). Each edge of the hexagon is necessarily equal to and parallel to its opposite edge. Such a hexagon, by a simple translation operation (altering its position on the plane without changing its orientation), will form a tiling pattern. The quadrilateral need not be convex. Exactly the same technique creates a tiling pattern for any nonconvex quadrilateral.

The case of the hexagon was settled in 1918 by K. Reinhardt in his doctoral thesis at the University of Frankfurt. He showed that any tessellating convex hexagon belongs to one of three classes. Kershner, in a 1969 article, "On Paving the Plane," explains the three types as follows.

Label the sides and angles of a hexagon as shown in Figure 79. A convex hexagon will tile the plane if and only if it belongs to one or more of the following classes:

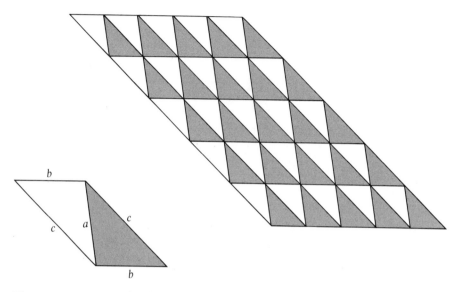

Figure 77 Any triangle tiles the plane

1. $A + B + C = 360°$,
    and $a = d$.
2. $A + B + D = 360°$,
    and $a = d$, $c = e$.
3. $A = C = E = 120°$,
    and $a = b$, $c = d$, $e = f$.

The illustration gives an example of each type of convex hexagon tiler and a portion of its tiling patterns. The gray lines outline a "fundamental region" that tiles the plane by translation. Note that Type 2 requires reflection if the hexagon is asymmetric.

In similar fashion the tessellating convex pentagons can be classified in eight ways. Five were found by Reinhardt. Kershner describes them by labeling the pentagon as shown in Figure 80. A convex pentagon paves the plane if it belongs to one or more of the following classes:

1. $A + B + C = 360°$.
2. $A + B + D = 360°$,
    and $a = d$.
3. $A = C = D = 120°$,
    and $a = b$, $d = c + e$.

4. $A = C = 90°$,
   and $a = b$, $c = d$.
5. $A = 60°$, $C = 120°$,
   and $a = b$, $c = d$.

Examples of each type and its tiling pattern are reproduced in the illustration with gray lines outlining fundamental regions. Only Type 2 requires reflection.

"At this point," writes Kershner, "either [Reinhardt's] technique or his fortitude failed him, and he closed the thesis with the statement that in principle it ought to be possible to complete the consideration of pentagons along the lines of his considerations up to that point, but it would be very tedious and there was always the possibility that no further types would emerge. Indeed, it is quite clear the Reinhardt and everyone else in the field thought that the Reinhardt pentagon list was probably complete.

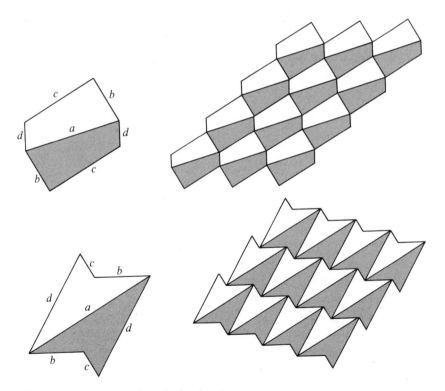

Figure 78 Any quadrilateral tiles the plane

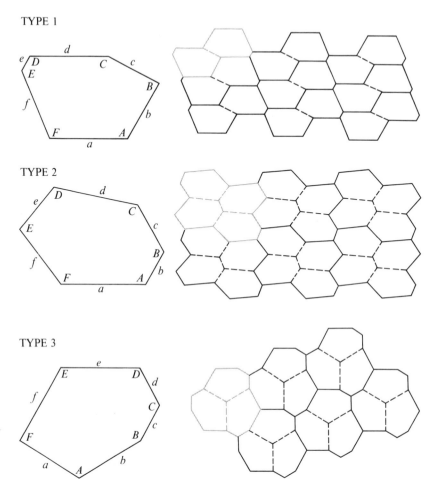

TYPE 1

TYPE 2

TYPE 3

Figure 79 The three types of convex hexagon tiler

"For reasons that I would have difficulty explaining I have been intrigued by this problem for some thirty-five years. Every five or ten years I have made some kind of attempt to solve the problem. Some two years ago I finally discovered a method of classifying the possibilities for pentagons in a more convenient way than Reinhardt's to yield an approach that was humanly possible to carry to completion (though just barely). The result of this investigation was the discovery that there were just three additional types of pentagon . . . that can pave the plane. These pavings are totally surprising. The discovery of their existence is a source of considerable gratification."

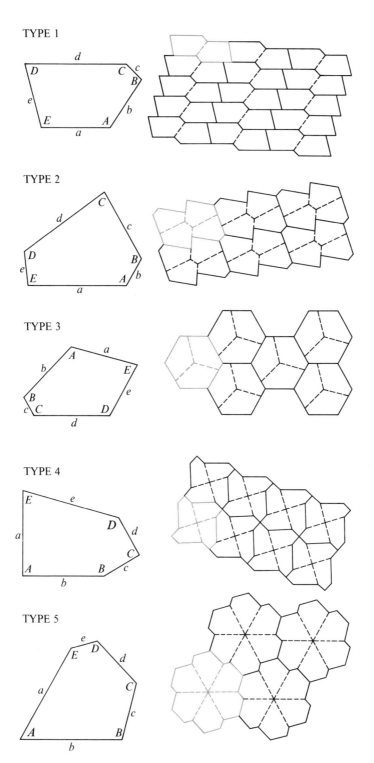

Figure 80 The five types of tiling convex pentagon known in 1918

The three additional types (*see* Figure 81) are described by Kershner as follows (Type 7 and Type 8 require reflection):

6. $A + B + D = 360°, A = 2C,$
   and $a = b = e, c = d.$
7. $2B + C = 2D + A = 360°,$
   and $a = b = c = d.$
8. $2A + B = 2D + C = 360°,$
   and $a = b = c = d.$

Kershner's paper does not include a proof that there are no other convex pentagons that tile the plane, "for the excellent reason," reads the editor's introductory note, "that a complete proof would require a rather large book."

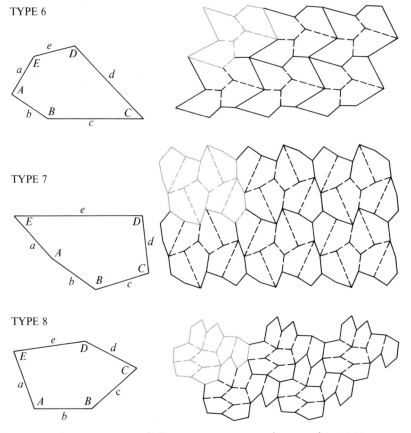

TYPE 6

TYPE 7

TYPE 8

Figure 81   Three new types of tiling convex pentagon discovered in 1967

Note that Kershner has deliberately drawn his tessellations with polygons that are as irregular as possible, within the limits of their type, in order to bring out the nature of the tessellation. The most regular hexagonal tessellation is, of course, the familiar beehive mosaic. One can readily see that it belongs to all three hexagon types.

If the beehive hexagons are bisected, the result is a pentagonal pattern that belongs to Type 1 (*see* Figure 82*A*). The pattern formed by six pentagons in a flowerlike arrangement (*see* Figure 82*B*) uses a tile that belongs to Type 1, Type 5, and Type 6. The most remarkable of all the pentagonal patterns is a tessellation of *equilateral* pentagons (*see* Figure 82*C*). It belongs to Types 2 and 4. Observe how quadruplets of these pentagons can be grouped into oblong hexagons in two different ways, each set tessellating the plane at right angles to the other. This beautiful tessellation is frequently seen as a street tiling in Cairo and occasionally in the mosaics of Moorish buildings. It underlies many of Escher's tessellations.

The equilateral pentagon is readily constructed with a compass and a straightedge (*see* Figure 83). First draw a side of the pentagon, *AB*. Construct its perpendicular bisector, *CD*, then draw lines *CE* and *CF* at 45 degrees to *AB*. With center at *A* and radius *AB*, draw a circular arc cutting *CE* at *P*. The same construction is repeated on the other side, with center at *B* and the arc cutting *CF* at *R*. Keeping the compass with radius *AB*, let *R* be the center and strike an arc that cuts the perpendicular bisector, *CD*, at *Q*.

The pentagon's corner angles at *P* and *R* are right angles. The corner at *Q* is a little more than 131 degrees, and corners *A* and *B* are a trifle more than 114 degrees. The length from *Q* to *B* is the product of a side of the pentagon and the square root of two. The pentagon's area (it is easy to prove) is precisely the square of line segment *CR*.

Among the infinite tessellations of the plane that can be made with congruent nonconvex polygons, combinatorial geometers have given special attention in recent years to tiling with polyominoes and their cousins the polyiamonds and polyhexes. (Polyominoes are formed by joining unit squares, polyiamonds by joining equilateral triangles, and polyhexes by joining regular hexagons.) Many fascinating problems have been formulated, some solved and some not. That will be the topic of the next chapter.

ADDENDUM

A remarkable letter on this chapter after it first appeared as a column in 1975, came from Richard E. James III, a computer scientist with the Control Data Corporation. He sent a strange tessellation (*see* Figure 84) along with a note

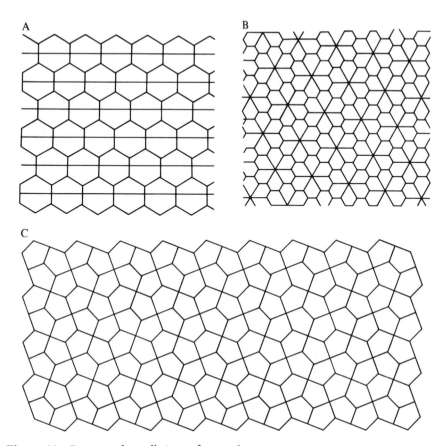

Figure 82   Pentagonal tessellations of unusual symmetry

describing the pentagon (in R. B. Kershner's notation) as $A = 90$ degrees, $C + D = 270$ degrees, $2D + E = 2C + B = 360$ degrees, and $a = b = c + e$. "Do you agree that Kershner missed this one?" he asked.

Kershner had indeed missed it. This means that the problem of classifying all convex pentagons that tile the plane is not solved, as Kershner had supposed. I must say that Kershner received the blow with grace and good humor. In a letter to him, I mentioned that James's discovery illustrates the pragmatic side of mathematical proof, namely that proofs are not known to be proofs until there is a consensus among experts. Kershner replied as follows:

"In connection with your philosophical comments on the nature of a proof you might be interested in an observation by that eminent authority, me. In *The Anatomy of Mathematics,* Kershner and L. R. Wilcox (Ronald Press, 1950) I wrote:

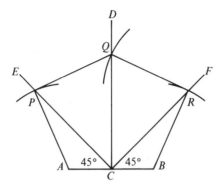

Figure 83 How to construct equilateral-pentagon tiler

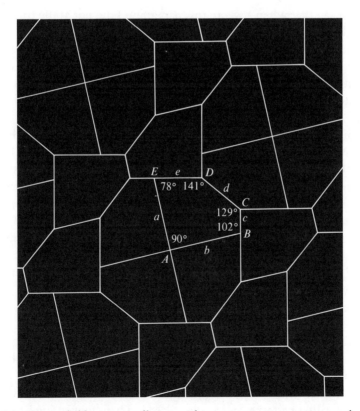

Figure 84 A remarkable new tessellation with congruent convex pentagons by Richard E. James III

Now it must be said that there is no simple test that can be applied to determine the validity of a proof, that is, to determine that an alleged proof really is a proof. Mathematical history contains rare instances of arguments that were generally accepted as proofs for hundreds of years, before being successfully challenged by a very ingenious mathematician, who pointed out a possibility that had been overlooked in the alleged proof. And more recently, every year there appear, in the mathematical journals of the world, a certain number of papers which point out that some statement, allegedly proved in a preceding paper, was not only erroneously proved (that is, not proved) but was, in fact, incorrect. These facts are mentioned for the benefit of those who feel that there is some magic formula for a proof which makes it immutable and unarguable henceforth and forevermore.

"I must say that when I wrote this paragraph I did not at the time propose eventually to illustrate its validity so graphically myself."

James's tessellation can be varied in ways that have been analyzed in a 1978 paper by Doris Schattschneider of Moravian College, Bethlehem, Pa. It is a basic pattern that could have been discovered by the medieval Moors or even by the ancient Greeks or Romans, but it is probable that the pattern had never before been seen by human eyes until James first put it on paper!

The discovery of new types of tiling pentagons did not end with James's finding. Marjorie Rice, a San Diego housewife with no mathematical training beyond the minimum required in high school, began a systematic search for new patterns. In 1976 she discovered a tenth type (*see* Figure 85), two more types later that year, and still another one the following year, bringing the total number of types to thirteen. A fourteenth type was found in 1985 by Rolf Stein, a mathematics graduate student at the University of Dortmund in West Germany. Its tiling pattern is on the cover of *Mathematics Magazine* (November 1985), with a note about it on page 308. As far as I know, no more new types have been discovered, although there is as yet no proof that the list is complete. Nor is there a full listing of all nonconvex pentagons that tile the plane.

Doris Schattschneider gives a brief account of Mrs. Rice's fantastic achievements in her 1978 paper, and a more detailed account in her contribution to *The Mathematical Gardner*. The latter paper includes three color plates of beautiful Escher-like tesselations (bees, fish, and flowers) that Mrs. Rice based on her new tiling patterns, and a color plate of a handwoven rug based on the James tesselation. The bee pattern provides the book's jacket.

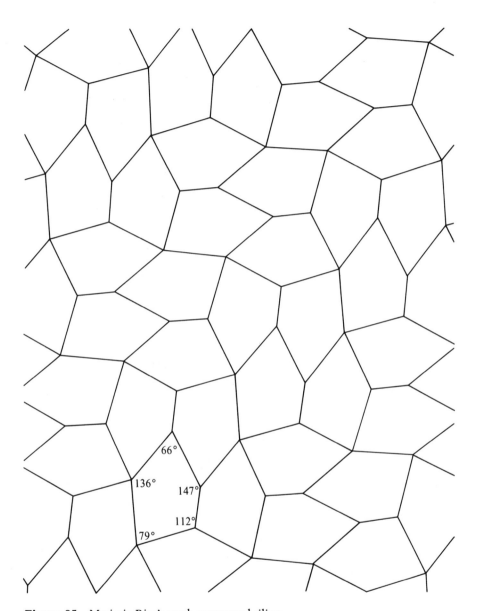

Figure 85 Marjorie Rice's tenth pentagonal tiling

B I B L I O G R A P H Y

"On Paving the Plane." R. B. Kershner, in *The American Mathematical Monthly* 75, October 1968, pp. 839–844.

"On Paving the Plane." R. B. Kershner, in *APL Technical Digest* 8, July–August 1969, pp. 4–10.

"The Laws of Sines and Cosines for Polygons." R. B. Kershner, in *Mathematics Magazine* 44, May 1971, pp. 150–153.

"Tesselations with Pentagons." J. A. Dunn, in *The Mathematical Gazette* 55, December 1971, pp. 366–369; see also 56, December 1972, pp. 332–335.

"Tesselations of Pentagons." Stanley R. Clemens, in *Mathematics Teaching* (a British journal for elementary teachers), June 1974, pp. 18–19. See follow-up by John Parker, *ibid.* March 1975, p. 34.

"Tiling the Plane with Congruent Pentagons." Doris Schattschneider, in *Mathematics Magazine* 51, January 1978, pp. 29–44. Reprinted in *Percy Alexander MacMahon: Collected Papers,* Vol. 2, George E. Andrews, ed. MIT Press, 1986.

"Convex Polygons that Cannot Tile the Plane." Ivan Niven, in *The American Mathematical Monthly* 85, December 1978, pp. 785–789.

"A Note on a Result of Niven on Impossible Tesselations." M. S. Klamkin and A. Liu, in *The American Mathematical Monthly* 87, October 1980, pp. 651–653.

"In Praise of Amateurs." Doris Schattschneider, in *The Mathematical Gardner,* David A. Klarner, ed. Prindle, Weber and Schmidt, 1981.

"Equilateral Convex Pentagons Which Tile the Plane." M. D. Hirschhorn and D. C. Hunt, in *Journal of Combinatorial Theory* 39, May 1985, pp. 1–18.

"Tilings by Polygons." Chapter 9 of *Tilings and Patterns.* Branko Grünbaum and Geoffrey C. Shephard. W. H. Freeman and Company, 1987. This marvelous 700-page work is the first comprehensive treatise on all aspects of tiling theory.

Tiling with Nonconvex Pentagons

"Tesselations with Equilateral Reflex Polygons." Gillian Hatch and David Simonds, in *Mathematics Teaching* September 1978, pp. 32–34.

"Tesselations of Non-convex Pentagons." Michael Palmer, *ibid,* March 1979, pp. 8–11.

"Spiral Tilings and Versatiles." Branko Grünbaum and G. C. Shephard, *ibid,* September 1979, pp. 50–51.

"The Incredible Pentagonal Versatile." Doris Schattschneider, designs by Marjorie Rice, *ibid,* December 1980, pp. 52–53.

Solution to Problem 67D, posed by H. Martyn Cundy, based on results by Doris Schattschneider, in *The Mathematical Gazette* 67, December 1983, pp. 307–309.

FOURTEEN

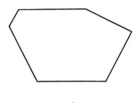

Tiling with Polyominoes, Polyiamonds, and Polyhexes

"I often wondered at my own mania
for making periodic drawings. . . .
What can be the reason of my being
alone in this field? Why does none
of my fellow-artists seem to be
fascinated as I am by these
interlocking shapes?"

—M. C. Escher

The previous chapter was devoted to tessellating the infinite plane with congruent, nonoverlapping convex polygons. In this chapter we extend the topic into the vast domain of nonconvex polygons, with special attention to polyominoes, polyiamonds and polyhexes.

Polyominoes are shapes formed by joining unit squares at their edges. They were first introduced to the mathematical world in 1953 by Solomon W. Golomb. His book *Polyominoes* (Scribner's, 1965) is the standard reference on this popular recreation. Some of my previous reports on polyominoes are reprinted in several earlier book collections of my *Scientific American* columns.

The monomino (single square) and domino obviously tile the plane, and so do the two kinds of tromino. It takes only a few moments to discover that each of the five tetrominoes will tile in a simple periodic pattern in which all the tiles are identically oriented; that is, no tile needs to be rotated or reflected (flipped over).

Each of the twelve five-celled figures will tile the plane. All but three — the *T*, *U*, and *R* pentominoes — tile by simple translation (sliding without rotations or reflections). Each of the three that require rotation can be fitted into pairs by turning one upside down to form an order-10 polyomino that tiles by translation (*see* Figure 86).

There are thirty-five distinct hexominoes, and all of them tile without being reflected. Some tile by simple translation. Those that do not can be joined in pairs by turning one upside down to make an order-12 polyomino that tiles by translation.

John H. Conway has long been interested in the tiling properties of polyominoes and other polygons. When he examined the 108 heptominoes, it was clear that the task of identifying the tilers would be tedious unless he could find a criterion that would, as he put it, "dispose of most of them rapidly without diagrams all over the graph paper."

The criterion discovered by Conway is an efficient one that applies to any polygon. It is based on a hexagonal tiling pattern (*see* diagrams at left and middle in Figure 87). Note that edges *a* and *d* are equal and parallel and that throughout the pattern *a* joins *d* with the two hexagons in the same orientation. Note also that each of the other four edges joins its corresponding edge on a tile that has been rotated 180 degrees.

With these facts in mind it is easy to understand the basis of Conway's criterion. We examine a given polygon to see if its perimeter can be divided into six parts, *a, b, c, d, e,* and *f,* that meet the following requirements:

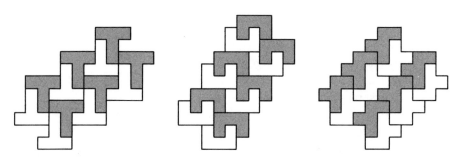

Figure 86 The three pentominoes that require rotation to tile periodically

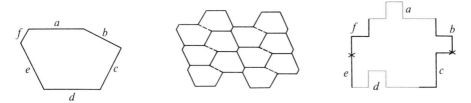

Figure 87   Criterion for periodic tiling without reflection

1. Two opposite edges, *a*, and *d*, are "parallel" in the sense that they are congruent and in the same orientation.
2. Each of the other four edges, *b, c, e,* and *f,* are centrosymmetric; that is, they are unaltered by a 180-degree rotation around a midpoint.

If the polygon meets both requirements, it will tile the plane periodically. No flipping over of tiles is necessary. The tiles are paired simply by turning one upside down to form a figure that tiles by translation.

Conway's example will make this procedure clear (*see* diagram at right in Figure 87). The two gray lines mark sides *a* and *d*, which are "parallel." The two X's distinguish the edges *b, c, e,* and *f.* Each of the four edges has central symmetry; therefore the figure will tile the plane by pairing. One figure of the pair is rotated 180 degrees with respect to the other, without reflection, and the double shape tiles by translation. It is important to realize, Conway adds, that any of the six edges may be empty (nonexistent). The criterion is quite general, applying to triangles, quadrilaterals, pentagons, and all higher polygons.

Of the 108 heptominoes, 101 meet Conway's criterion. That means each can be paired with an upside-down mate to form a 14-cell shape that tiles by translation. The seven that fail to meet the criterion are shown in Figure 88. The first heptomino obviously cannot tile the plane because there is no way to fill the hole. The fourth one tiles by pairing with a 90-degree rotation as is shown in Figure 89, (lower left).

The fifth, which Conway considers the most interesting heptomino, tiles in three ways. It tiles by pairing, with 90-degree rotation and reflection. Without reflection it tiles in quadruplets of four orientations (*see* Figure 89, (top right, shows on of two ways).

The second heptomino in this group will not tile without reflection. The smallest region, which tiles by translation, contains four replicas in four orientations; two of the replicas are reflected, as is shown in Figure 89, (lower right). This is the lowest-order polyomino, unique among the heptominoes, that requires reflection to tile the plane.

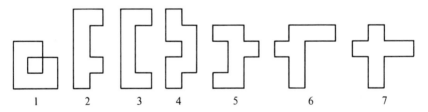

Figure 88 The seven heptominoes that do not meet Conway's criterion for tiling periodically

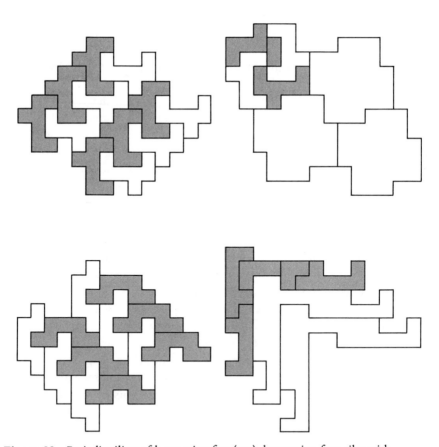

Figure 89 Periodic tiling of heptomino five (top), heptomino four tiles with 90-degree rotation (bottom left), and heptomino two tiles only with reflection (bottom right)

The third, sixth, and seventh heptominoes will not tile, making four non-tilers in all. Proving impossibility in each case is not difficult, although it is sometimes tedious. First you try all possible ways of fitting two replicas together, eliminating the ways that have a hole or a space that cannot be filled by a third replica. After all such pairs have been found, you test all ways of adding a third piece that allow the placing of a fourth. Eventually you reach a set of n heptominoes such that, no matter how they are combined, there is no way to add another replica. I leave these proofs to interested readers.

Although the third heptomino will not tile the plane, it can be combined with 3-by-3 square tiles to form a striking tessellation. The reader is urged to make a set of cardboard replicas of this heptomino, together with replicas of the order-3 square, and see if he can combine the shapes to form a tessellation.

As far as I know, only David Bird of North Shields, England, has isolated the nontilers among the 369 octominoes. The six with holes can of course be eliminated immediately. Bird reports twenty other nontilers, making twenty-six in all (*see* Figure 90).

Figure 90 The twenty-six nontiling octominoes

Is there a general algorithm that can be applied to any polyomino to decide if it will tile? Conway's criterion identifies certain polyominoes as tilers, but if a polyomino does not meet his criterion, it may or may not tile in other ways. Conway conjectures that there is no general algorithm, but this conjecture has not yet been proved. A major step toward establishing the undecidability of the algorithm was reported by Golomb in his 1970 paper "Tiling with Sets of Polyominoes." In that paper Golomb considers whether there is a procedure for deciding if any given finite set of different polyominoes (assuming an unlimited supply of each kind) will tile the plane. He shows that the problem is equivalent to a problem of tiling the plane with a finite set of edge-colored squares by matching colors at the joined edges. Since Hao Wang and his colleagues had earlier proved the latter problem to be undecidable (*see* "Games, Logic and Computers," by Hao Wang; *Scientific American*, November 1965) the corresponding problem for polyominoes is also undecidable.

Golomb's paper extended the results published in his 1966 article "Tiling with Polyominoes." In that paper he considered the tiling of such subsets of the plane as the half-plane, the quarter-plane, straight strips, bent strips, and rectangles. A chart summarizes results through the hexominoes.

If there is no region in a tessellation that tiles by translation, the tessellation is said to be nonperiodic. One curious way of tiling nonperiodically with congruent polygons is to use a set of them to build a larger replica. Obviously sets of these larger replicas can then be combined in the same way to make a still larger replica, and in that way the entire plane can be covered. A tile with the property of self-replication was named by Golomb a "rep-tile" [See his paper "Replicating Figures in the Plane," and the chapter on rep-tiles in my *Unexpected Hanging* (Simon & Schuster, 1969).]

Less work has been done on polyiamonds (figures formed by joining congruent equilateral triangles) than on polyominoes. (On the polyiamonds see Golomb's book and Chapters 18 and 24 in my *Sixth Book of Mathematical Games* (University of Chicago Press, 1983). It is not hard to establish by Conway's criterion that all polyiamonds through order 6 will tile. Of the twenty-four heptiamonds only the *V* heptiamond is not a tiler. (A simple impossibility proof is given in Chapter 24 of my *Sixth Book*). Bird, Gregory J. Bishop, Andrew L. Clarke, John W. Harris, Wade Philpott, and others have found that all octiamonds will tile. As far as I know, the 160 enneiamonds (order 9) have not yet been settled, although two correspondents, Bird and Clarke, agree that the nontilers are the twenty-one shown in Figure 91.

The polyhexes, formed by joining congruent regular hexagons (*see* the chapter on them in my *Mathematical Magic Show*, Knopf, 1971), have received

Figure 91 The twenty-one enneiamonds believed to be nontilers

even less attention than the polyiamonds. Bird and Bishop have established that all polyhexes through order 5 are plane-fillers. Of the eighty-two hexahexes Bird believes only five are nontilers (*see* Figure 92).

Once it is determined that a given polygon will tile, one can ask in how many distinct ways the tiling can be accomplished. It can be a very sticky question. A. W. Bell, in his paper "Tessellations of Polyominoes," has made a beginning by attempting to classify all patterns for the *L* tetromino. He obtained nineteen patterns but made no claim for completeness.

Major P. A. MacMahon's little book *New Mathematical Pastimes* (Cambridge University Press, 1921; why has no publisher reprinted it?) explains with a wealth of illustrations how a simple polygon tessellation is easily transformed into a more complicated one. You merely take a pair of straight edges, which

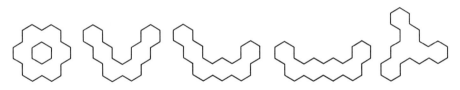

Figure 92 The five hexahexes thought to be nontilers

always go together in the pattern, and change the straight boundary to a crooked one (curves are allowed), subject to certain symmetry restrictions. It was by just such techniques that Islamic craftsmen created the intricate abstract mosaics in the Alhambra and the Taj Mahal.

Here is how Escher described the difficulty of adapting the technique to the creation of tiles that resemble living things: "The borderline between two adjacent shapes having a double function, the act of tracing such a line is a complicated business. On either side of it, simultaneously, a recognizability takes shape. But the human eye and mind cannot be busy with two things at the same moment, and so there must be a quick and continual jumping from one side to the other."

Tessellation theory, quite apart from its usefulness to artists who design patterns for walls, floors, fabrics, and so on, has a practical application in industry. In cutting congruent shapes from thin sheets of metal, plastic, cardboard, leather, and other materials, a tessellation pattern obviously provides the only way of doing it without waste. An unusual art book could be made by collecting pictures of the tessellations used in modern manufacturing, from the simple rectangular patterns of postage stamps, dollar bills, and playing cards to complicated machine parts.

There is also a potential application to jigsaw puzzles. In 1958 the British mathematical physicist Roger Penrose amused himself by transforming a parallelogram tessellation into tilings of order-18 polyiamonds. Some of them make excellent puzzles. Consider a "loaded wheelbarrow" polyiamond (*see* Figure 93). If the reader will make twelve cardboard replicas of the figure, he will find it a challenging task to fit them together to make a region that tiles the plane by translation.

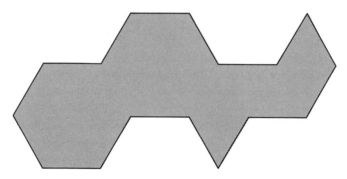

Figure 93 Roger Penrose's polyiamond puzzle

Penrose is Rouse Ball Professor of Mathematics at the Mathematical Institute of the University of Oxford. He is best known among physicists for his contributions to relativity theory and cosmology. He and his father, L. S. Penrose, were the first to discover "impossible objects" such as the famous Penrose staircase that Escher used so effectively in his lithograph *Ascending and Descending.*

For about a decade it has been known that there are sets of polygons that together will not tile the plane periodically but will do so nonperiodically. A few years ago Raphael M. Robinson constructed a set of six tiles that tile only nonperiodically. Penrose later found a set of four and finally a set of just two. Is there a single shape, duplicates of which will tile only nonperiodically? This is one of the deepest unsolved problems in tessellation theory. The subject of nonperiodic tiling is, however, another story.

ANSWERS

One way of tesselating the plane with replicas of a certain heptomino and a 3-by-3 square is shown in Figure 94. There are other ways to do it.

A tiling pattern for Roger Penrose's wheelbarrow is shown in Figure 95. Replicas of the hexagonal region (consisting of twelve polyiamonds in twelve different orientations) will tile the plane periodically in the manner of the familiar beehive pattern of regular hexagons. Although the tiling pattern is unique, there are many different ways a fundamental region of twelve pieces can be outlined on the infinite plane.

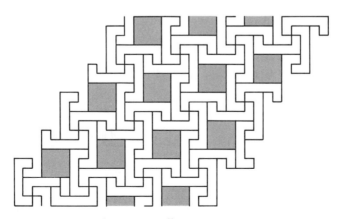

Figure 94 A heptomino-and-square tessellation

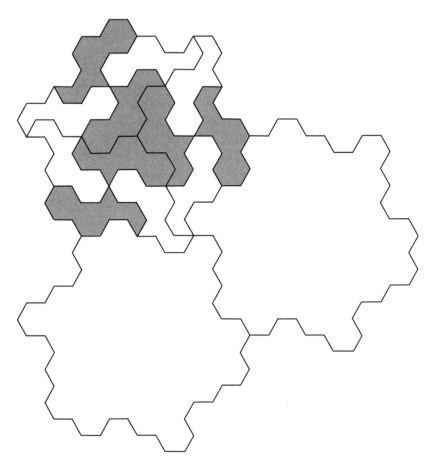

Figure 95 How the Penrose polyiamond tiles the plane

B I B L I O G R A P H Y

"Replicating Figures in the Plane." S. W. Golomb, in *Mathematical Gazette* 48, December 1964, pp. 403–412.

"Tiling with Polyominoes." S. W. Golomb, in *Journal of Combinatorial Theory* 1, September 1966, pp. 280–296.

"Tesselations of Polyominoes." A. W. Bell, in *Mathematical Reflections,* edited by members of the Association of Teachers of Mathematics, Cambridge University Press, 1970.

"Tiling with Sets of Polyominoes." S. W. Golomb, in *Journal of Combinatorial Theory* 9, July 1970, pp. 60–71.

"Will it Tile? Try the Conway Criterion!" Doris Schattschneider, in *Mathematics Magazine* 53, September 1980, pp. 224–233.

"Tiling with Congruent Tiles." Branko Grünbaum and G. C. Shephard, in *Bulletin of the American Mathematical Society* (new series) 3, November 1980, pp. 951–973.

"Polymorphic Polyominoes." Anne Fontaine and George E. Martin, in *Mathematics Magazine* 57, November 1984, pp. 275–283.

"Tilings by Polygons." Chapter 9 of *Tilings and Patterns*. Branko Grünbaum and Geoffrey C. Shephard. W. H. Freeman and Company, 1987.

FIFTEEN

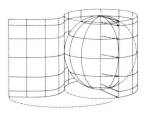

.
.
.

Curious Maps

"I quite realized," said Columbus,
"That the Earth was not a rhombus,
But I *am* a little annoyed
To find it an oblate spheroid."

—E. CLERIHEW BENTLEY

If only the earth were flat, as Martin Luther, John Calvin, and the fathers of the Catholic Church believed, the labors of cartographers would be greatly simplified. Indeed, flat-earth maps of the first six centuries of the Christian Era presented no serious geometrical problems. A few learned churchmen agreed with Pythagoras, Plato, Aristotle, and Archimedes that the earth was round, but to most churchmen such a belief was heresy.

Early medieval maps of the world were "Scripture-preserving." They were either rectangular to preserve the "four corners" in Isaiah 11:12 and in Revelation 7:1, or they were circular or oval to preserve the "circle of the earth" in Isaiah 40:22. There was, of course, no need for meridians and parallels. Jerusalem was exactly in the center, as Ezekiel 5:5 suggests. The top of the map pointed east and included the site of Eden. The land was surrounded by the "great waters" that had once flooded the earth and also by the sources of the "four winds" (Daniel 7:2, Revelation 7:1) that blew so erratically toward the Holy City.

After the eighth century the rotundity of the earth gradually became acceptable to the Church, with such eminent Catholics as Thomas Aquinas and Dante Alighieri defending it. It slipped out of favor among the Reformers, but during the Renaissance it quickly won the day. The rapid increase in travel and exploration, particularly the great sea voyages, made it necessary to have better maps, and that naturally revived a troublesome mathematical question: How can a portion of the earth's surface be drawn on a plane so that all distances are accurately represented?

The answer is that it cannot. The side of a cylinder or cone will map perfectly onto a plane, but the surface of a sphere will not. You can flatten a cylinder or cone without distorting its surface, but even a small region of a sphere's surface will not press flat without cracking, folding, or stretching. Every flat map of all or part of the earth distorts something. The cartographer's tricky task is to design maps that will show the least distortion or no distortion of those properties the map's user deems desirable. At the same time, the distortion of all other properties should be minimal. We shall take a quick look at some classical methods of mapmaking before turning to methods that result in more bizarre maps.

One of the most desirable features of a map is that angles between any two lines on the map be the same as the angles between those same lines on the globe. This feature is enormously useful at sea because it means that observed angles between two landmarks correspond to angles measured on the map with a protractor, and also because small regions on such a map preserve their shape. Maps of that type are called conformal. The simplest way to produce a conformal map is by "stereographic projection." As Figure 96 shows, a sphere is projected by straight lines from point B on the sphere's surface to a plane tangent to the sphere at a point opposite. The projection is called equatorial, polar, or oblique, depending respectively on whether the antipodes are on the equator, the poles, or somewhere else. The price paid for conformality is a distortion of the scale factor that increases with distance from the center of the map.

If the projection to the tangent plane is made from the globe's center, it is a gnomonic projection, so called because it is related to the construction of a sundial with a gnomon. Every great circle on the globe becomes a straight line on a gnomonic map. The map is not conformal, but for navigators it has one merit that all other planar projections lack. A straight line between any two

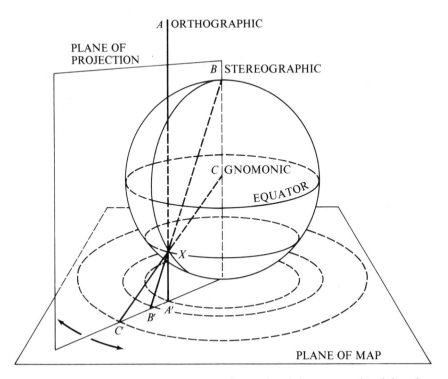

Figure 96 Three azimuthal projections: orthographic *(A)*, stereographic *(B)*, and gnomonic *(C)*

points on a gnomonic map corresponds to a great circle on the globe and therefore provides the geodesic: the shortest distance between the two points.

Since the point of projection can be at any spot inside or outside the globe, there is an endless variety of perspective projections. If the point is at infinity (all projecting lines parallel), the projection is orthographic. Our view of the moon or a view of the earth from the moon is essentially orthographic. Distance distortions are great at the edges of an orthographic map. The map preserves neither area nor angles, but if it is skillfully drawn, it gives a strong illusion of the earth's roundness. Perspective maps with the "eye" above the earth may be among the least accurate with respect to many properties, but they are the most accurate in matching our visual perceptions of a sphere.

Projections need not be made onto a plane. They can be made onto surrounding cylinders or cones that can then be cut and unrolled. Imagine the earth snugly fitted inside a cylinder. The projecting lines are parallel to the plane that cuts the great circle where the globe and the cylinder touch (*see* Figure 97). The resulting map has the amazing property, highly desirable for

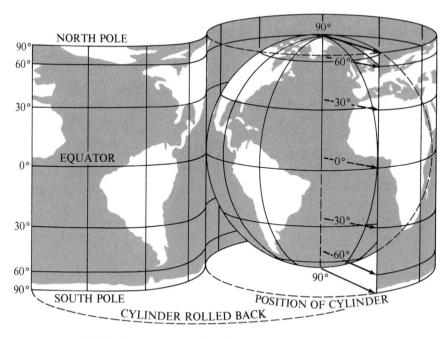

Figure 97   Cylindrical projection method for making an equal-area map

many purposes, that areas are preserved: All closed curves have the same areas as their corresponding curves on the globe, and they have them in scale. If the cylinder touches the earth along the Equator, all meridians and parallels on the map become straight lines meeting at right angles.

The equal-area cylindrical map is not conformal, and it severely distorts shapes and distances. Indeed, it is not hard to prove that no map can simultaneously be conformal and area-preserving. A great variety of other area-preserving maps have been devised. In modern atlases one of the most popular of equqal-area world maps is an elliptical projection worked out by Karl B. Mollweide in 1805.

The cylindrical projection suggested to Gerhardus Mercator, a sixteenth-century Flemish geographer, the famous conformal map that bears his name. Imagine the earth's surface punctured at the poles, the two holes enlarged until the surface is a cylinder, the cylinder stretched along its length until the map is conformal, then cut along a meridian and unrolled. There is an enormous distortion of scale near the poles. As we learned in grade school, this familiar world map shows Greenland larger than South America when actually it is much smaller. (In order to minimize such scale variations, modern atlases use a modification of the Mercator map called the Miller projection.)

The Mercator projection has, however, one remarkable property that makes it invaluable to navigators. If you rule a straight line between any two points on the map, the line is a loxodrome, or rhumb line, connecting the two points. A loxodrome is a line that keeps a constant angle with parallels and meridians (*see* Figure 98). Imagine a point on the globe that starts at the Equator and moves northward in any constant compass direction. The path will be a loxodrome that spirals toward the North Pole, finally strangling the pole after an infinity of turns around it. On a stereographic map (with its plane tangent to the North Pole) a loxodrome projects as a logarithmic spiral.

The loxodrome is not the shortest distance between two points, but for small distances it is reasonably close to a geodesic, and it has the practical value of being a path that does not require constant changes of bearing. For long distances, navigators usually determine the geodesic from a gnomonic projection, then break it into shorter rhumb lines on a Mercator map to minimize changes in compass settings.

So much for the classic projections. Let us turn now to more radical distortions. For a two-point equidistant projection, points *A* and *B* are selected. The map is then drawn so that all distances from *A* and *B* to any other point on the map are in true scale. Such a map is useful to a person traveling from *A* to *B*. No matter how circuitous one's route is, one can at any time measure with a ruler

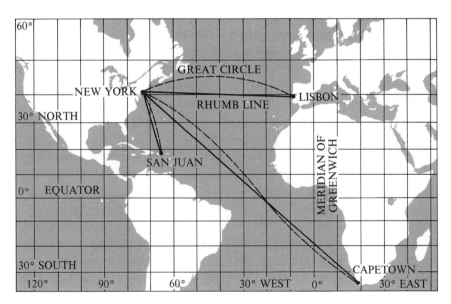

Figure 98 A Mercator conformal map with loxodromes, or rhumb lines, from New York

exactly how far one is from both points. Suppose the two points on a two-point equidistant map of the world are the two poles. What will the map look like?

Another curious special-purpose map is the "Mecca map," designed to show at once to a Moslem the exact direction he must face when he prays at any spot on the globe. One way to draw such a map is to make a stereographic projection with Mecca at the plane's tangent point. Because the map is conformal, Mecca's bearing angle can be determined by measuring the angle between a straight line to Mecca and a meridian. Unfortunately, the meridians on such a map are curved, making it difficult to measure the angle exactly. One can, however, construct a Mecca map on which all meridians are straight lines, making it possible to measure bearing angles with a protractor. Such a map, with Mecca replaced by another holy place, Wall Street, is given in an internal Bell Laboratories memorandum on map oddities written by Edgar N. Gilbert, a mathematician (*see* Figure 99). The map's upper boundary is the North Pole.

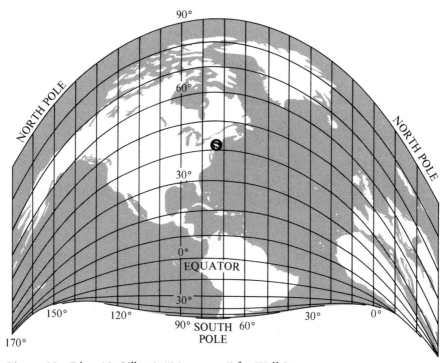

Figure 99 Edgar N. Gilbert's "Mecca map" for Wall Street

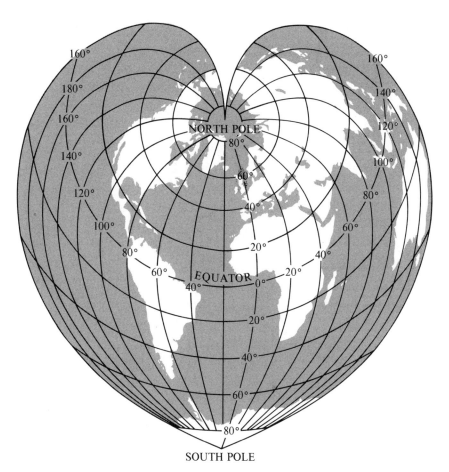

Figure 100 Johann Werner's cordiform (heart-shaped) equal-area map

Gilbert's memorandum contains even stranger maps. One is a cordiform (heart-shaped) equal-area map invented by Johann Werner (*see* Figure 100). It was popular in the sixteenth century but, writes Gilbert, has "now fallen into undeserved obscurity. The highly distorted parts of the map lie away from most major land masses. The curved latitude lines give the map a pleasing illusion of roundness. . . . The latitude lines are arcs of circles, evenly spaced and centered at the North Pole. The longitude lines are drawn to make distances along the latitude lines the same in the map as in the sphere."

The bunching of land masses on Werner's map reflects an actual bunching of land on the earth. The Pacific Ocean is so immense that if you look at a globe from a point above the English Channel, you will see more than 80 percent of the world's land, and the hemisphere opposite will be almost entirely water. Is such lopsidedness surprising? Gilbert came up with an answer by replacing the continents with small nonoverlapping circular caps. Assuming that N such caps are randomly distributed around the globe, what is the probability that the centers of all N circles lie in one hemisphere? In a paper titled "The Probability of Covering a Sphere with N Circular Caps," Gilbert shows that the probability is $2^{-N}(N^2 - N + 2)$. With $N = 7$ continents the probability is $11/32$, proving that the earth's lopsided land distribution is not at all remarkable. The reader may enjoy testing the formula by determining the probability that the centers of one, two, or three continents all lie in one hemisphere.

The strangest of all Gilbert's maps was made by taking a conformal world map based on a conical projection and projecting it back onto a sphere to produce a conformal "two-world map." Figure 101 shows how the globe looks in perspective from a point about five radii from its center. When people visit Gilbert's office, he likes to ask them what is wrong with his globe. If the visitor cannot see what is wrong, Gilbert give the globe one slow, complete turn. "Even this hint," he writes, "does not always succeed." Actually every spot on the globe has a duplicate on the other side! Unless you are an experienced geographer, however, it is not easy to realize that you are seeing much more of the world than can normally be seen on one hemisphere.

With the aid of computer-generated graphics it is now possible to write programs that distort a map so that the areas express some desired value such as annual rainfall, retail sales, and so on. Regions on the map still retain recognizable shapes in spite of the distortions. The joke map showing a New Yorker's idea of the United States, although it goes back (in many variants) to precomputer days, is a familiar example of such maps. One of the top experts on such special-purpose map projections is Waldo R. Tobler, a geographer at the University of Michigan. In a paper titled "A Continuous Transformation Useful in Districting," he explains a computer program that distorts a map to show relative populations of regions by their relative sizes, and he shows how such a technique could be a valuable aid in planning voting districts. In 1973 a geographer with a fondness for floppy bow ties was given a presentation award that consisted of a framed world map distorted to the shape of his tie. Drafting the map was no problem for Tobler's computer program.

Dissection maps (as I call them) are world maps projected on a pattern of squares, triangles, or polygonal tiles of other shapes. The tiles can be fitted together to make an "interrupted map" (a map with discontinuities) of any portion of the globe. The philosopher and mathematician Charles Sanders Peirce designed such a conformal map. It is a projection on eight isoceles right

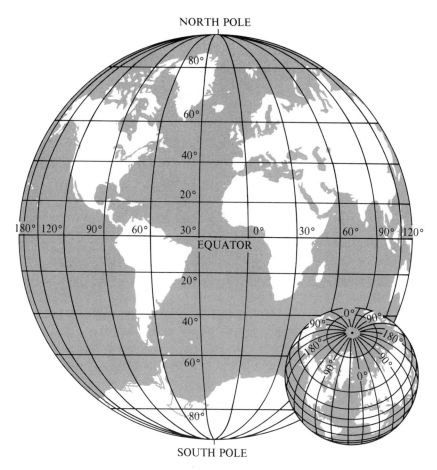

Figure 101 Gilbert's "two-world" globe and an oblique view of the globe

triangles that may be regarded as the faces of an octahedron that has been flattened until a space diagonal is zero. The vertexes of the zero diagonal are the North and South poles of Peirce's map.

B. J. S. Cahill of Oakland, Calif. patented his butterfly map in 1913, and it enjoyed a considerable vogue in the 1930s. The world is projected onto the eight equilateral-triangular faces of a regular octahedron. Cahill had several versions of his map, based on different projections, but they all consisted of eight triangular tiles that could be fitted together as one pleased (*see* Figure 102).

R. Buckminster Fuller's first Dymaxion map was a projection of the world onto the fourteen faces (six squares and eight equilateral triangles) of a cuboc-

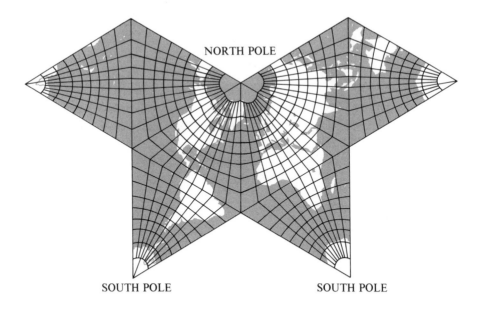

NORTH POLE

SOUTH POLE   SOUTH POLE

Figure 102   Butterfly map by B. J. S. Cahill

tahedron. Gerard Piel, who was then science editor of *Life,* was so intrigued by it that he asked the cartographer Richard Edes Harrison to draw an unfolded net of the solid. Staff artists completed a color map of the drawing, and it was published on cover-stock paper in the March 1, 1943, issue of *Life.* It was a great success. All over the country, in homes and laboratories, one would see the little cuboctahedron hanging on a cord and rotating with currents of air.

At about the same time, Irving Fisher, a distinguished Yale economist, thought of a similar idea: a gnomonic projection of the world to the twenty triangular faces of an icosahedron. Figure 103 shows an unfolded net of this Platonic solid as it appears inside the jacket of *World Maps and Globes,* an entertaining introduction to cartography by Fisher and O. M. Miller. Harrison was the map's cartographer.

The reader may enjoy pasting a copy of Fisher's gnomonic projection on heavy paper and folding it into the icosahedral globe or cutting out the triangles and fitting them together in different ways. If the poles are put at opposite vertexes, the Equator becomes a straight line. Fisher marketed several versions of his Likaglobe, as he called it, and wrote an article about its merits in *Geographical Review* (October 1943). The globe can, of course, be projected on other regular solids, but distortion is exaggerated on the cube and the tetrahedron. Although the dodecahedron makes a handsome pseudoglobe, its faces cannot be used as tiles because regular pentagons will not tessellate the plane.

The icosahedron seems to be the ideal polyhedron for a dissection map, and Fuller himself has now adopted it. In 1954 he copyrighted his Dymaxion Skyocean Projection World Map, drawn by Shoji Sadao. It differs from Fisher's Likaglobe in having the North and South poles on opposite faces at points slightly off center. For a small fee, readers can obtain a punch-out version printed in four colors on heavy stock. It folds into a beautiful icosahedron that rests on a cardboard stand that comes with the map. (Orders should go to Dymaxion Maps, 3500 Market Street, Philadelphia, PA, 19104). Also available are several wall-map versions of the unfolded icosahedron and an explanation sheet on which Fuller discusses the philosophy and mechanics of the projection.

We have considered only a small portion of the many curious special-purpose maps designed by ingenious cartographers. Harrison once drew a world map consisting of nothing but ocean shipping lanes. From a distance, you could see the continents clearly, but from close up you discovered that the map contained not a single shoreline. At the Christian Science Publishing House in Boston, you can step inside a world globe 30 feet in diameter. If such a globe were transparent and viewed from the outside, everything would be mirror-reversed. From the inside the land masses and oceans appear normal.

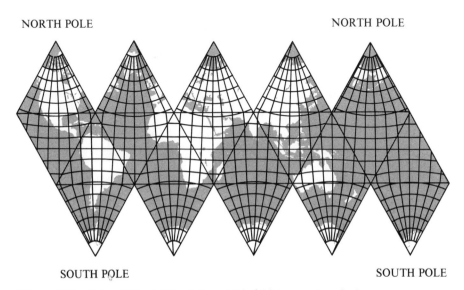

NORTH POLE   NORTH POLE

SOUTH POLE   SOUTH POLE

Figure 103 Irving Fisher's Likaglobe, which folds into an icosahedron

The major tenet of a bizarre little sect called Koreshanity — it flourished in America late in the nineteenth century — was that we actually live on the undersurface of such a hollow earth, with the entire cosmos inverted and compressed to fill the interior. The sect was founded by a baptist named Cyrus Reed Teed, of Utica, N.Y. You will find Teed's inside-out maps in his great scientific work *The Cellular Cosmogony* (1870) and in the pages of *The Flaming Sword,* a periodical that the cult kept going until 1949. How did Teed explain the fact that we cannot point a telescope straight up and see the earth's other side? Well, he took care of it with special optical laws describing how light travels in bent paths. Teed's cosmos has not been sufficiently appreciated by philosophers of science. By performing an inversion operation on the universe, then inventing new laws of physics, one can obtain an inside-out cosmology that is not easy to refute except by slashing it with Occam's razor.

A large class of eccentric maps that we have not considered are the maps of imaginary regions. "Might-have-been" maps show what the world might have looked like if major wars had ended differently. Fantasy maps depict Oz, Hell, Eden, Poictesme, Narnia, Barsoom, Middle-Earth, Atlantis, and other fanciful realms. J. B. Post has collected ninety-eight of them in his beautiful *Atlas of Fantasy.* And we must not forget the Bellman's ocean chart in Fit 2 of Lewis Carroll's *Hunting of the Snark:*

> He had bought a large map representing
> the sea,
> Without the least vestige of land:
> And the crew were much pleased when
> they found it to be
> A map they could all understand
>
> "What's the good of Mercator's North
> Poles and Equators,
> Tropics, Zones, and Meridian Lines?"
> So the Bellman would cry: and the crew
> would reply,
> "They are merely conventional signs!
>
> "Other maps are such shapes, with their
> islands and capes!
> But we've got our brave Captain
> to thank"
> (So the crew would protest) "that he's
> bought *us* the best —
> A perfect and absolute blank!"

Fisher's book on world maps closes with thirty-six excellent questions and answers about world geography. How well can the reader do (without consulting a map) on the following selection?

1. You are on a ship five miles from an entrance to the Panama Canal and sailing due west toward it. In what body of water is your ship?
2. Flying due south from Detroit, what foreign country do you reach first?
3. Which is nearer Miami, California or Brazil?
4. Which is farther north, Venice or Halifax?
5. Which is farther south, Venice or Vladivostok?
6. Which is larger, Japan or Great Britain?
7. What four states in the U.S. touch at one point?
8. Does a geodesic (great circle) from Tokyo to the Panama Canal pass east or west of San Francisco?

The illustrations for this chapter are by Richard Edes Harrison. He also gave me invaluable assistance on the text.

ANSWERS

A two-point equidistant map of the world, when the two points are any pair of antipodes, is a straight line. The next problem dealt with the distribution of continents around the world. If one, two, or three nonoverlapping circular caps are randomly distributed around the globe, the probability that their centers lie on one hemisphere (any half sphere, not necessarily the Northern or the Southern Hemisphere) is 1.

The answers to the geographical questions are:

1. Pacific.
2. Canada.
3. California.
4. Venice.
5. Vladivostok.
6. Japan.
7. Arizona, Colorado, New Mexico, Utah.
8. East.

With reference to question 7, I learned in 1984 that the Four Corners Highway passes within a few hundred yards of the spot where the four states meet. An access road leads to the spot, where you find local Indians selling wares to tourists.

ADDENDUM

So many strange projections of world maps came to my attention after I wrote this chapter that I have space only to comment briefly on a few. I had mentioned the New Yorker's map of the United States on which New York, California, and Florida are huge, all other states tiny. In the late 1970s, the Danish writer and inventor Piet Hein marketed in Denmark what he called the Denmark Globe, on which Denmark is enlarged to symbolize how big it seems to Danes.

In 1985 I clipped an advertisement from *The New York Review of Books* (February 14) for a map of the Americas with South America on top, and the United States and Canada at the bottom. Of course all place names are right-side up. "Run of the mill maps place the U.S. on top," the ad read. "Since 'upper' is equated with 'superior,' this breeds misconceptions and mischief. The Turnabout Map offers a corrective perspective."

Professor Tobler, cited in the chapter, wrote to me about his unpublished Möbius-strip map. The earth is projected onto the strip, like a Mercator projection that has been stretched horizontally, twisted, and the ends joined. The two stretched polar points become the single edge of the one-sided surface. The map has the following remarkable property: A pin puncturing it at any point emerges at the spherical antipodal point! Who but a topologist, Tobler asked, would suspect this similarity of a sphere and a Möbius surface?

B I B L I O G R A P H Y

"Quincuncial Projections of the Sphere." Charles S. Peirce, in *American Journal of Mathematics* 2, 1879, pp. 394–396.

World Maps and Globes. Irving Fisher and O. M. Miller. Essential Books, 1944.

Elements of Cartography. A. H. Robinson. Wiley, 1960.

"The Probability of Covering a Sphere with *N* Circular Caps." E. N. Gilbert, in *Biometrika* 52, 1965, p. 323.

Notes and Comments on the Composition of Terrestrial and Celestial Maps. J. H. Lambert. Translated and introduced by W. R. Tobler. University of Michigan Publication 8, 1972.

Atlas of Fantasy. J. B. Post. Mirage Press, 1973.

"A Continuous Transformation Useful in Districting." W. R. Tobler, in *Annals of the New York Academy of Sciences* 119, 1973, pp. 215–220.

"Distortion in Maps." E. N. Gilbert, in *SIAM Review* 16, January 1974, pp. 47–62.

Cartographical Curiosities. Gillian Hill. British Museum Publications, 1978.

SIXTEEN

.
.
.

The Sixth Symbol and Other Problems

1. WHAT SYMBOL COMES NEXT?

Can you sketch the sixth figure in the sequence of five symbols shown in Figure 104?

2. WHICH SYMBOL IS DIFFERENT?

Every person who ever took an I.Q. test has surely been annoyed by questions that involve a row of symbols and a request to identify the symbol that "does not belong" to the set. Sometimes there are so many different ways a symbol can be different from the others that brighter students are penalized by having to waste time deciding which symbol is "most obviously different" to the person who designed the test.

It was irritation over such ambiguity that prompted Tom Ransom, a puzzle expert of Toronto, Ontario, to devise a delightful parody of such a test. The reader is asked to inspect the five symbols in Figure 105 and pick out the one that is "most different."

Figure 104   What is the sixth figure in this series?

3. CUTTING A CAKE

A cake has been baked in the form of a rectangular parallelepiped with a square base. Assume that the square cake is frosted on the top and four sides and that the frosting's thickness is negligible (zero). We want to cut the cake into *n* pieces so that each piece has the same volume and the same area of frosting. The slicing is conventional. Seen from above, the cuts are like spokes radiating from the square's center, and each cutting plane is perpendicular to the cake's base.

If *n* is 2, 4, or 8, the problem is easily solved by slicing the cake into two, four, or eight congruent solids. Suppose, however, that *n* is 7. How can we locate the required seven points on the perimeter of the cake's top? If you solve it for 7, you will be able to generalize to any *n*.

This pretty problem is given by H. S. M. Coxeter on page 38 of his classic *Introduction to Geometry* (Wiley, 1967). Coxeter does not answer it, but it should give readers little difficulty. The general solution is surprisingly simple.

4. TWO CRYPTARITHMS

Here are two elegant cryptarithms by Alan Wayne that have not been published before. The first is in French, the second in German. Each letter represents just one decimal digit, and we adopt the usual convention that zero must not begin a number. Both have unique solutions.

Figure 105   Tom Ransom's I.Q. test: Which symbol is the "most different"?

$$
\begin{array}{c}
\ \ \text{V I N G T} \\
+\ \ \ \ \ \ \text{C I N Q} \\
\underline{\ \ \ \ \text{C I N Q}} \\
\ \ \text{T R E N T E}
\end{array}
\qquad
\begin{array}{c}
\ \text{E I N} \\
+\ \text{E I N} \\
\ \text{E I N} \\
\underline{\ \text{E I N}} \\
\ \text{V I E R}
\end{array}
$$

5. LEWIS CARROLL'S "SONNET"

In 1887 Lewis Carroll included in a letter to Maud Standen, a "child-friend," a six-line poem that he called an "anagrammatic" sonnet. (He was using "sonnet" in an older sense, meaning any short piece of verse.) "Each line has four feet," Carroll wrote to Maud, "and each foot is an anagram, *i.e.,* the letters of it can be rearranged so as to make one word." Most of the anagrams, he said, had been devised "for some delicious children" he had met the previous summer at Eastbourne. The words vary in length from four through seven letters, and it is assumed that proper names are not allowed.

Here is the "sonnet":

> As to the war, try elm. I tried.
> The wig cast in, I went to ride
> 'Ring? Yes.' We rang. 'Let's rap.'
>   We don't.
> 'O shew her wit!' As yet she won't.
> Saw eel in Rome. Dry one: he's wet.
> I am dry. O forge! Th' rogue! Why
>   a net?

In most cases there is little doubt about the correct word. For example, the first foot, "As to," could be "oast" or "stoa," but more likely Carroll meant the commoner word "oats." For several of the feet, however, the intended word is not clear. No solution by Carroll has survived, and to this day there is contention among Carrollians over the precise set of twenty-four words Carroll had in mind. Readers are invited to make their own list to compare with the conjectures in the answer section.

6. THIRD-MAN THEME

Two kings are the sole occupants of a chessboard (*see* Figure 106). The task is to add a third man, creating a position that meets these provisos:

1. Neither king is in check.
2. The position can be reached in a legal game.
3. It can be proved, by retrograde analysis of previous legal moves, that neither side has a legal play.

Note carefully the wording. It asks not for a double stalemate but only for a position in which neither side can move. The solution is unique.

This sophisticated little problem appeared in *The Problemist* (September 1974, p. 471), where it was credited to G. Husserl of Israel. Newman Guttman called it to my attention. The original problem asked for the minimum number of men that must be added to the board to meet the conditions, but I have made the problem easier by stating that the minimum is one.

ANSWERS

1. The sixth figure is given in Figure 107, with the vertical lines of symmetry shown in gray. On the right of each gray line are the numerals 1 through 6; on the left are their mirror reflections.

2. The first symbol of Tom Ransom's I.Q. test differs from the other four in having a gray border. The second symbol differs in not being shaded. The

Figure 106 Where and what is the third man?

Figure 107   Solution to the sixth-symbol problem

third differs in not being square, and the fourth in having a gray spot. The fifth is the only symbol that is not unique with respect to some property, and therefore it is the one symbol "most different" from the others. To put it differently, the first four are each different in the same general way, whereas the fifth is different in a different way, which makes it radically different.

One is reminded of the story about the report of a psychiatrist on the most remarkable case in his experience. The patient was a person who showed no trace of any kind of neurosis. In my opinion Ransom has found a simple, elegant model of a semantic paradox that is frequently encountered but seldom explicitly recognized.

3.   To cut a square cake, frosted on the top and four sides, into *n* pieces of equal volume and equal frosting area, we need only mark the perimeter into *n* equal parts and cut the cake in the usual manner (*see* Figure 108). To understand why this is so, we adopt a stratagem given by Norman N. Nelson and Forest N. Fisch in their article "The Classical Cake Problem" (*The Mathematics Teacher* 66, November 1973, pp. 659–661). Imagine the cake cut along its diagonals into four congruent triangular prisms, and the prisms arranged in a row as shown at the right in the illustration. The base line is the cake's perimeter. Divide the cake into seven equal portions, *A, B, C, D, E, F, G,* as shown by the seven gray lines. Those lines correspond to the seven cutting planes in the solution.

It is easy to see that the area of icing on each portion is the same. The sums of the one or two rectangles that form the sides of each portion clearly are equal.

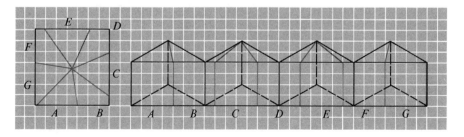

Figure 108   Solution to the square-cake problem

The tops of each portion (composed of one or two triangles) are also equal in area because the sums of the bases of these triangles are equal, and all triangles have the same altitude. Finally, the volumes of the seven portions are the same because their heights are equal and their tops are equal in area.

The solution obviously generalizes to n portions. We cannot, however, apply the procedure to a cake on which the frosting has a finite thickness, because the four corners of such a cake introduce complexities.

Stephen I. Warshaw of the Lawrence Livermore Laboratory of the University of California generalized the cake-cutting problem. The procedure given for the square cake, he showed, also applies to any cake with a polygonal base each side of which is tangent to the same circle. This includes all triangles and rhombuses, all regular polygons (as well as the limiting case of the circle), and all polygons of unequal sides that meet the proviso. Even more surprising, Warshaw found that this generalization led to the solving of a sticky problem in his work: "How to find a simple way to divide a discrete mesh cell into parts of given proportions in computer simulation studies involving hydrodynamic motions. The solution, once seen, is 'obvious.'"

4. The French cryptarithm is solved with $94,851 + 6,483 + 6,483 = 107,817$, and the German one with $821 + 821 + 821 + 821 = 3284$.

Many readers pointed out that the numeral 1 is commonly called *eins* in German, not *ein* as it appeared in the German cryptarithm. William C. Giessen was the first to add that if this change is made, the cryptarithm still has a unique solution: $1329 + 1329 + 1329 + 1329 = 5316$.

5. The first published attempt to solve Lewis Carroll's anagram poem was made by Sidney H. Williams and Falconer Madan in their *Handbook of the Literature of the Rev. C. L. Dodgson* (Oxford University Press, 1931). Some of their errors were corrected in a revised edition, the *Lewis Carroll Handbook* (Oxford University Press, 1962), by Roger L. Green. For further criticism and speculation see Philip S. Benham, "Sonnet Illuminate," in *Jabberwocky*, the journal of the Lewis Carroll Society (Summer 1974).

Drawing on the above sources, and with help from Spencer D. Brown, here is how matters stand on the words Carroll most likely had in mind.

1. As to: *oats* (not *oast, stoa*).
2. The war: *wreath* (not *thawer*).
3. Try elm: *myrtle*.
4. I tried: *tidier*.
5. The wig: *weight*.
6. Cast in: *antics* (not *sciant, actins,* or *nastic*).
7. I went: *twine*.
8. To ride: *editor* or *rioted*.

9. Ring yes: *syringe.*
10. We rang: *gnawer.*
11. Let's rap: *plaster* (not *stapler, persalt, palters, psalter,* or *platers*).
12. We don't: *wonted.*
13. O shew: *whose.*
14. Her wit: *writhe* (not *wither, whiter*).
15. As yet: *yeast.*
16. She won't: *snoweth.*
17. Saw eel: *weasel.*
18. In Rome: *merino* (not *moiren, minore*).
19. Dry one: *yonder.*
20. He's wet: *seweth* (not *thewes, hewest*).
21. I am dry: *myriad.*
22. O forge: *forego.*
23. Th' rogue: *tougher* (not *rougeth*).
24. Why a net: *yawneth.*

Carroll's letter containing the puzzle poem first appeared in *Six Letters by Lewis Carroll,* privately printed in 1924. It is reprinted as Letter XLV in *A Selection from the Letters of Lewis Carroll to His Child-Friends,* edited by Evelyn M. Hatch (Folcroft, 1973).

Now for some hot news concerning number 16 above. Can you rearrange the letters of "she won't" to make another common two-word phrase?

6. The only solution to the third-man chess problem is to place a white bishop on the board as shown in Figure 109. Black obviously cannot move,

Figure 109 Solution to the third-man theme

and White cannot move because it is not his move. To prove that it must be Black's move (therefore Black is stalemated) requires elementary retrograde analysis that I leave to the reader.

Many readers could not comprehend why the third-man chess problem does not have many other solutions, such as a white pawn in place of the white bishop. The reason is that none of these positions rules out the possibility that it is White's move, and if White can move, the position fails to meet the demand that neither side can move. For instance, assume that the black king is on the QB1 cell. It is checked by an advancing white pawn. The black king moves to the corner. Now it is White's move. The given solution is the only one for which it can be shown by retrograde analysis that it must be Black's move, proving that the position is a stalemate for Black.

SEVENTEEN

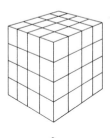

Magic Squares and Cubes

"In my younger days, having once
some leisure (which I still think I
might have employed more usefully),
I had amused myself in
making . . . magic squares."

—BENJAMIN FRANKLIN, from a letter

Two major breakthroughs have been made in the study of magic squares and magic cubes: All order-5 magic squares have been counted, and the first perfect magic cube has been constructed. I am pleased to be the first to publish both results. So that the magnitude of the two achievements can be fully appreciated, let us take a brief look at the history of magic squares.

Although some of the greatest mathematicians have done work on magic squares, and even though such work leads into the theories of groups, lattices, Latin squares, determinants, partitions, matrices, congruence arithmetic, and other nontrivial areas of mathematics, the most enthusiastic square constructors have been amateurs. The famous Franklin square, an ingenious 16-by-16 matrix that Benjamin Franklin called "the most magically magical of any magic square ever made by any magician," is itself the topic of many articles and monographs. The literature on magic squares in general is vast, most of it written by laymen who had become hooked on the elegant symmetries of these interlocking number patterns.

A standard magic square, as few readers need to be told, is a square array of positive integers from 1 through N^2, arranged so that the sum of every row, every column, and each of the two main diagonals is the same. N is the "order" of the square. It is easy to see that the magic constant is the sum of all the numbers divided by N. The formula is

$$(1 + 2 + 3 + \cdots + N^2)/N = \tfrac{1}{2}(N^3 + N)$$

The trivial square of order 1 is simply the number 1, and of course it is unique. It is equally trivial to prove that no order-2 square is possible.

There are eight ways to arrange the digits 1 through 9 in an order-3 array that is magic. It is traditional, however, not to count rotations and reflections. When they are excluded, the order-3 square is unique. To appreciate the gem-like beauty of this most ancient of all combinatorial curiosities, consider all the ways that its constant, 15, can be partitioned into a triplet of distinct positive integers. There are exactly eight:

.

$$9 + 5 + 1$$
$$9 + 4 + 2$$
$$8 + 6 + 1$$
$$8 + 5 + 2$$
$$8 + 4 + 3$$
$$7 + 6 + 2$$
$$7 + 5 + 3$$
$$6 + 5 + 4$$

Now, in the order-3 square each of eight lines of three numbers must total 15: the six orthogonals (rows and columns) and the two diagonals. The eight lines exactly match the number of triplets we have available. Since the center number belongs to a row, a column and both diagonals, it clearly must be a digit that appears in four of the eight triplets. The only such digit is 5. We therefore know that 5 is the central number.

Consider 9. It belongs to only two triplets. We cannot place it in a corner since each corner cell belongs to three lines. Consequently it must go in a side cell. Because of the square's symmetry, it does not matter which side cell we choose, so let us put it above the 5. For the top corners, on each side of 9, we have no choices except 2 and 4. Again it does not matter which digit goes where, since one arrangement is merely a mirror reflection of the other. The rest of the square follows automatically. We have by this simple construction proved its uniqueness.

The completed square, in the form shown in Figure 110, is the *Lo shu* of ancient China. According to legend the pattern was first revealed on the shell of a sacred turtle that crawled out of the Lo River in the twenty-third century B.C., but today's Chinese scholars trace references to it back no further than the fourth century B.C. From then until the tenth century the pattern was a mystical Chinese symbol of enormous significance. The even numbers were identified with yin, the female principle, and the odd numbers with yang, the male principle. The central 5 represented the earth, around which, in evenly balanced yin and yang, were the other four elements: 4 and 9 symbolizing metal, 2 and 7 fire, 1 and 6 water, and 3 and 8 wood.

There are 880 magic squares (excluding rotations and reflections) of order 4. They were first given by Bernard Frénicle de Bessy in 1693. There are many ways to classify them. One of the best was devised by Henry Ernest Dudeney, who explains his system in an excellent article on magic squares in early printings of the fourteenth edition of the *Encyclopaedia Britannica*. The last printing of that edition substitutes for Dudeney's article a superb historical article by Schuyler Cammann. The current (fifteenth) edition has a trivial microarticle on magic squares in the *Micropaedia*.

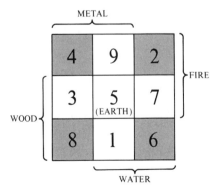

Figure 110 The *Lo shu* magic square of ancient China

How many magic squares are of order 5? The best estimate was given by Albert L. Candy in his *Construction, Classification and Census of Magic Squares of Order Five,* privately published in Lincoln, Neb., in 1938. Candy arrived at a total of 13,288,952. The exact number was not known until 1973, when the counting was completed by a computer program developed by Richard Schroeppel, a mathematician and computer programmer at Information International. The program, using a standard backtracking procedure, consists of about 3500 "words" and took about 100 hours of running time on a PDP-10. A final report, written by Michael Beeler, was issued in October 1975.

Candy's estimate was low by a wide margin. Not counting rotations and reflections, there are 275,305,224 magic squares of order 5. Schroeppel prefers to divide that number by 4 and give the total as 68,826,306. The reason is that in addition to the eight variants obtained by rotation and reflection, there are four other variants generated by the following two transformations, which also preserve magic:

1. Exchange the left and right border columns, then exchange the top and bottom border rows.
2. Exchange rows 1 and 2 and rows 4 and 5. Then exchange columns 1 and 2 and columns 4 and 5.

When these two transformations are combined with the two reflections and four rotations, the result is $2 \times 4 \times 2 \times 2 = 32$ forms that can be called isomorphic. With this definition of isomorphic the count becomes 68,826,306.

That number can be lowered even more by considering another well-known transformation. If every number in a magic square is subtracted from $N^2 + 1$ (in this case 26), the result, called the complement, is also magic. When the center of an order-5 square is 13, the complement is isomorphic with the original. If it is not 13, a different square results. If we broaden the term isomorphic to include complements, the count of order-5 squares drops to about 35 million.

The task of classifying order-5 squares in meaningful ways is staggering. Dudeney once wrote that certain ways of dividing magic squares into types seemed to him about as useful as dividing people into those who take snuff and those who do not. Nevertheless, certain divisions yield unexpected results. Consider, for example, the total number of order-5 squares with centers that are numbers 1 through 13:

1. 1,091,448
2. 1,366,179
3. 1,914,984
4. 1,958,837
5. 2,431,806
6. 2,600,879
7. 3,016,881
8. 3,112,161
9. 3,472,540
10. 3,344,034
11. 3,933,818
12. 3,784,618
13. 4,769,936

Note that the totals steadily increase from 1 through 8 but that the totals for 9 and 10 and for 11 and 12 are reversed. That there are more squares with a center of 11 than squares with a center of 12, and more squares with a center of 9 than squares with a center of 10, came as a surprise. Of course, these same anomalies occur in the counts of squares with centers 14 through 25, since every square with a center that is not 13 has its complement. There are as many squares with 1 in the center as there are with 25, and the same is true for all numbers except 13.

Figure 111 shows an order-5 square of a type that is more powerfully magic than any other. It is associative, which means that any pair of numbers sym-

1	15	24	8	17
23	7	16	5	14
20	4	13	22	6
12	21	10	19	3
9	18	2	11	25

Figure 111 A pandiagonal, associative magic square of order 5

metrically opposite the center add up to $N^2 + 1$. And it is pandiagonal (sometimes called Nasik or diabolic), which means that its broken diagonals add up to 65, the constant. To put it another way, if we tile the plane with this square, we can outline a 5-by-5 square anywhere on this infinite pattern and it will be magic, although not necessarily associative (*see* Figure 112). To be associative too, it must have 13 in the center.

The *Lo shu* is associative but not pandiagonal. An order-4 square may be pandiagonal or associative but not both. The order-5 square is the smallest one that can have both properties. Excluding rotations and reflections, 3600 order-5 squares are pandiagonal, or if we also exclude variants obtained by the cyclic permutation of rows and columns, 144 are pandiagonal. In other words, there are 144 infinite patterns of the type shown here, each containing 25 pandiagonal order-5 squares. Of the 144, just 16 contain a square that is also associative. All of this, by the way, was known before Schroeppel developed his computer program.

Of the sixteen associative pandiagonal squares of order 5, four have 1 in the first cell, four have 1 in the third cell, four have 1 in the seventh and four have 1 in the eighth. The medieval Moslems were particularly intrigued by pandiagonal squares with 1 in the center. The patterns were not, of course, associa-

1	15	24	8	17	1	15	24	8	17	1	15	24	8	17
23	7	16	5	14	23	7	16	5	14	23	7	16	5	14
20	4	13	22	6	20	4	13	22	6	20	4	13	22	6
12	21	10	19	3	12	21	10	19	3	12	21	10	19	3
9	18	2	11	25	9	18	2	11	25	9	18	2	11	25
1	15	24	8	17	1	15	24	8	17	1	15	24	8	17
23	7	16	5	14	23	7	16	5	14	23	7	16	5	14
20	4	13	22	6	20	4	13	22	6	20	4	13	22	6
12	21	10	19	3	12	21	10	19	3	12	21	10	19	3
9	18	2	11	25	9	18	2	11	25	9	18	2	11	25

Figure 112   Cyclic permutations of the order-5 square

tive, but the Moslems thought of the central 1 as being symbolic of the unity of Allah. Indeed, they were so awed by that symbol that they often left blank the central cell on which 1 was supposed to go.

It is natural to extend the concept of magic squares to three dimensions and even higher ones. A perfect magic cube is a cubical array of positive integers from 1 to N^3 such that every straight line of N cells adds up to a constant. These lines include the orthogonals (the lines parallel to an edge), the two main diagonals of every orthogonal cross section and the four space diagonals. The constant is

$$(1 + 2 + 3 + \cdots + N^3)/N^2 = \tfrac{1}{2}(N^4 + N)$$

There is, of course, a unique perfect cube of order 1, and it is trivially true that there is none of order 2. Is there one of order 3? Unfortunately, 3 does not quite make it. I do not know who first proved the impossibility, but Richard Lewis Myers, Jr., has a simple way of doing it. Consider any 3-by-3 cross section. Let A, B, C be the numbers of the first row, D, E, F the numbers of the third, and X the central number. Since the diagonals and the center column each must add up to 42, we can write

$$3X + A + B + C + D + E + F = 3(42)$$

From this we subtract $A + B + C + D + E + F = 2(42)$ to get $3X = 42$, and $X = 14$. Since 14 cannot be the center of every cross section, the cube is impossible.

Annoyed by the refusal of such a cube to exist, magic-cube buffs have relaxed the requirements to define a species of semiperfect cube that apparently does exist in all orders higher than 2. These are cubes where only the orthogonals and four space diagonals are magic. Let us call them Andrews cubes, since W. S. Andrews devotes two chapters to them in his pioneering *Magic Squares and Cubes* (1917). The order-3 Andrews cube must be associative, with 14 in its center. There are four such cubes, not counting rotations and reflections. All are given by Andrews, although he seems not to have realized that they exhaust all basic types.

No perfect cube of order 4 exists. As far as I know, the first proof was published by Schroeppel in "Artificial Intelligence Memo 239," (MIT, 1972). The first step is to show that on any 4-by-4 section (orthogonal or diagonal) the four corners must add up to the constant. Let Q be the constant, and label the sixteen cells with other letters (*see* Figure 113). The gray lines indicate six quadruplets that catch all sixteen cells. Since each corner cell is common to

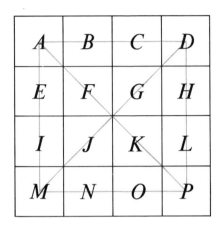

Figure 113 Richard Schroeppel's proof, lemma 1

three lines, $3A + 3D + 3M + 3P$ plus each of the other cells taken once must equal $6Q$. If we subtract from this the values of the four rows, we are left with $2A + 2D + 2M + 2P = 2Q$, which reduces to $A + D + M + P = Q$, our first lemma.

Now consider the cube's eight corners. We prove that any two corners connected by an edge must have a sum of $Q/2$. Call the corners A and B. Let C, D and E, F be the corners of any two edges parallel to A, B (*see* Figure 114). $ABDC$, $EFBA$, and $EFDC$ are each the corners of a 4-by-4 cross section, so that their total is $3Q$. Gather the like terms:

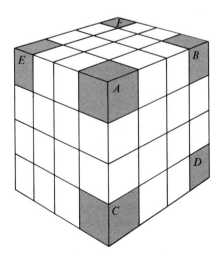

Figure 114 Richard Schroeppel's proof, lemma 2

$$2A + 2B + 2C + 2D + 2E + 2F = 3Q$$

Divide each side by 2:

$$A + B + C + D + E + F = 3Q/2$$

From this we subtract $C + D + E + F = Q$ to obtain $A + B = Q/2$, our second lemma.

Now consider corner B. It is joined to corners A, D, F. Since $A + B = F + B = D + B$, we can take B from each equality to prove that $A = F = D$. That is impossible, and so our proof is complete.

Is there a perfect magic cube of order 5? No one knows. Schroeppel has made a beginning by proving (using algebra and combinatorial thinking) that if such a cube exists, its center must be 63.

There are perfect magic cubes of order 8. A method of constructing them by the millions was discovered in the spring of 1970 by Myers, when he was sixteen and a student at William Tennant High School in Johnsville, Pa. He sent me a short note about it, saying that he had obtained his first cube "after three months, seven theories, and thirty-one sheets of graph paper." I am embarrassed to confess that I did not appreciate the significance of his claim. He did not send an actual cube, and I replied by suggesting a mathematical journal that could evaluate his work.

I next heard about Myers's cubes in December 1972 from John H. Staib, a mathematician at Drexel University in Philadelphia, where Myers had enrolled as a freshman. Staib sent an order-8 cube (*see* Figure 115), and although he extolled its symmetries and provided hints about how Myers had constructed it (by superposing three Latin cubes and applying base-8 notation), I still failed to comprehend the cube's importance. I had glanced too quickly at Andrew's book, noting references to magic cubes of order 3 and higher, without realizing that those cubes were only semiperfect. It was not until I started researching this report that I awoke to a full realization of what Myers had done.

Every orthogonal and diagonal line of eight numbers on the Myers cube shown in Figure 115, including the four space diagonals, add up to 2052. The cube is associative: Any two numbers symmetrically opposite the center add up to 513. It follows that not only do the eight corner cells of the cube total 2052 but also (as Staib pointed out) the corners of every rectangular solid centered in the cube add up to the same number. If that is not enough, the cube can be sliced into sixty-four order-2 cubes, and the eight numbers in each of these cubes add up to the constant!

1

```
19  497 255 285 432 78  324 162
303 205 451 33  148 370 128 414
336 174 420 66  243 273 31  509
116 402 160 382 463 45  291 193
486 8   266 236 89  443 181 343
218 316 54  472 357 135 393 107
185 347 85  439 262 232 490 12
389 103 361 139 58  476 214 312
```

5

```
381 159 401 115 194 292 46  464
65  419 173 335 510 32  274 244
34  452 206 304 413 127 369 147
286 256 498 20  161 323 77  431
140 362 104 390 311 213 475 57
440 86  348 186 11  489 231 261
471 53  315 217 108 394 136 358
235 265 7   485 344 182 444 90
```

2

```
134 360 106 396 313 219 469 55
442 92  342 184 5   487 233 267
473 59  30  215 102 392 138 364
229 263 9   491 346 188 438 88
371 145 415 125 208 302 36  450
79  429 163 321 500 18  288 254
48  462 196 290 403 113 383 157
276 242 512 30  175 333 67  417
```

6

```
492 10  264 230 87  437 187 345
216 310 60  474 363 137 391 101
183 341 91  441 268 234 488 6
395 105 359 133 56  470 220 314
29  511 241 275 418 68  334 176
289 195 461 47  158 384 114 404
322 164 430 80  253 287 17  499
126 416 146 372 449 35  301 207
```

3

```
306 212 478 64  141 367 97  387
14  496 226 260 433 83  349 191
109 399 129 355 466 52  318 224
337 179 445 95  238 272 2   484
199 293 43  457 380 154 408 118
507 25  279 245 72  422 172 330
412 122 376 150 39  453 203 297
168 326 76  426 283 249 503 21
```

7

```
96  446 180 338 483 1   271 237
356 130 400 110 223 317 51  465
259 225 495 13  192 350 84  434
63  477 211 305 388 98  368 142
425 75  325 167 22  504 250 284
149 375 121 411 298 204 454 40
246 280 26  508 329 171 421 71
458 44  294 200 117 407 153 379
```

4

```
423 69  331 169 28  506 248 278
155 377 119 405 296 198 460 42
252 282 24  502 327 165 427 73
456 38  300 202 123 409 151 373
82  436 190 352 493 15  257 227
366 144 386 100 209 307 61  479
269 239 481 3   178 340 94  448
49  467 221 319 398 112 354 132
```

8

```
201 299 37  455 374 152 410 124
501 23  281 251 74  428 166 328
406 120 378 156 41  459 197 295
170 332 70  424 277 247 505 27
320 222 468 50  131 353 111 397
4   482 240 270 447 93  339 177
99  385 143 365 480 62  308 210
351 189 435 81  228 258 16  494
```

Figure 115 Cross section of Richard Lewis Myers's magic cube of order 8 [Computer printout courtesy of William Gosper]

These remarkable symmetries make possible an enormous number of rearrangements of the cube, all in a sense isomorphic, and of course every arrangement can be rotated and reflected forty-eight ways. Imagine this cube with each of its 512 cells replaced by the same cube in any of its thousands of rearrangements or orientations. In cell 1 we put a cube that starts with 1. In cell 2 we put a cube that starts with $8^3 + 1 = 513$, in cell 3 we put a cube that starts with $(2 \times 8^3) + 1 = 1025$, and so on for the other cells. The result is a perfect magic cube of order 64. Order-64 cubes in turn will build perfect magic cubes of order 512, and the same is true for all higher orders that are powers of 8.

How many order-8 magic cubes are there? By choosing different Latin cubes to superpose, Myers can construct millions different from one another and from the one given here, although not all will be associative. The number of Latin squares of order 8 have been counted (there are billions), but the number of Latin cubes has not, so that the problem of counting just the order-8 cubes that can be generated by Myers's procedure alone is horrendous.

Is 8 the lowest order a perfect magic cube can have? Are there magic cubes of orders not in the 8-power series? Both are open questions.

ADDENDUM

When I wrote the foregoing text I believed that Richard Myers was the first to construct a perfect magic cube of order 8. As I quickly learned (of course, this in no way detracts from Myers's achievement), this is not the case.

Victor Meally called my attention to the construction of a perfect magic cube of order 8 in the late 1930s by J. Barkley Rosser and Robert J. Walker. It is pandiagonal in the sense that if any of its three sets of parallel sections are cyclically permuted, the cube remains perfectly magic. The cube was not published, but its construction is given in a manuscript deposited in a library at Cornell University. In the manuscript, the authors show that perfect pandiagonal cubes exist for all orders that are multiples of 8 and all odd orders higher than 8. A brief summary of the construction technique for the order-8 cube is given in W. W. Rouse Ball's *Mathematical Recreations and Essays* (Dover, 1987), revised by H. S. M. Coxeter.

Rosser and Walker were not, however, the first to construct a perfect magic cube of order 8. James G. Mauldon, an Amherst College mathematician, wrote to me about a remarkable 1888 paper by F. A. P. Barnard. He credits the first such cube to an unknown person who apparently published it in *The Commercial,* a Cincinnati newspaper, on March 11, 1875. (Details can be found in the fifth chapter of *Magic Cubes* by William Benson and Oswald

Jacoby.) Mauldon also sent his construction of a perfect associative magic cube of order 7, and perfect associative pandiagonal magic cubes of orders 9 and 11.

The first perfect pandiagonal cube of order 8 seems to have been constructed by C. Planck, who reported it in his rare, little-known *Theory of Path Nasiks* (privately published in Rugby, England, 1905). Planck showed that in k-space, where k is 2 or higher, the smallest perfect pandiagonal "cube" is of order 2^k, and the smallest that is also associative (symmetrically opposite pairs of numbers have the same sum) is $2^k + 1$. We saw how that is true of magic squares, where $k = 2$. In 3-space the smallest perfect pandiagonal cube is of order 8, and the smallest that is also associative is of order 9. Myers's cube is associative but not pandiagonal.

The first publication known to me of a perfect magic cube of order 7 is in *Play Mathematics* by Harry Langman (Hafner, 1962, pp. 75–76). After this chapter was published as a column (January 1976), scores of readers sent me such order-7 cubes, as well as construction procedures for orders 9, 11, and higher. Readers also explored magic cubes in dimensions above 3. For a listing of articles on magic cubes and hypercubes that have appeared in the *Journal of Recreational Mathematics*, see the bibliography of John Hendricks's 1985 paper.

As far as I know, no one has yet found a perfect magic cube of orders 5, 6, or 10, or proved that such cubes cannot exist.

The 880 magic squares of order 4 are depicted in the 1976 book on magic squares by Benson and Jacoby, along with considerable original material that appears in print for the first time. The book gives the fantastic tri-magic square of order 32 constructed by Captain Benson in 1949. This is a square that retains its magic not only when every number is squared but also when every number is cubed. The simplest-known square of that type was of order 64, before Captain Benson found a way to halve the order.

B I B L I O G R A P H Y

"Theory of Magic Squares and Cubes." F. A. P. Barnard, in *The Memoirs of the National Academy of Sciences* 4, 1888, pp. 209–270.

Magic Squares and Cubes. W. S. Andrews. Open Court, 1917. Dover reprint, 1960.

"The Algebraic Theory of Diabolic Magic Squares." J. Barkley Rosser and Robert J. Walker, in *Duke Mathematical Journal* 5, December 1939, pp. 705–728.

"Magic Squares." Martin Gardner, in *The Second Scientific American Book of Mathematical Puzzles and Diversions.* Simon and Schuster, 1961.

Magic Squares. John L. Fults. Open Court, 1974.

"Perfect Magic Cubes of Order 7." B. E. Wynne, in *Journal of Recreational Mathematics* 8, 1975–1976, pp. 285–293.

New Recreations with Magic Squares. William H. Benson and Oswald Jacoby. Dover, 1976.

"Magic Cubes and Prouhet Sequences." Allan Adler and Shuo-Yen R. Li, in *The American Mathematical Monthly* 84, October 1977, pp. 618–627.

Magic Cubes. William H. Benson and Oswald Jacoby. Dover, 1981.

The Wonders of Magic Squares. Jim Moran. Vintage, 1982.

Researches in Magic Squares. Akira Hirayama and Gakuho Abe. Osaka Kyoikutosho Co., Osaka, Japan, 1983. This 316-page book is so filled with new material on magic squares, cubes, and related constructions that it deserves an English translation.

"Ten Magic Tessaracts of Order Three." John R. Hendricks, in *Journal of Recreational Mathematics* 18, 1985–1986, pp. 125–134.

EIGHTEEN

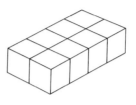

. . .

Block Packing

> "Pack my box with five dozen liquor jugs."
>
> — Anonymous pangram

A pangram is an attempt to pack as many different letters as possible into the shortest intelligible sentence. It is considered not cricket to use names and initials, such as "Schwartz" and "X. Q. Zym," or strange words, such as "pyrzqxgl," which in *The Magic of Oz* enables you to change instantly into any kind of animal you like if you know how to pronounce it correctly.

Many ultimate pangrams of twenty-six different letters have been constructed by word players, but they tend to be inelegantly obscure, for example "Vext cwm fly zing jabs kurd qoph," by Dmitri Borgmann. It means that an annoyed fly in a Welsh mountain hollow, humming shrilly, pokes at the nineteenth letter of the Hebrew alphabet drawn by a Kurd. Cryptographers find a perfect pangram amusing because it can be written in a simple letter-substitution cipher as ABCDE FGHIJ KLMNO PQRST UVWXY Z.

The difference between creating pangrams and working on packing problems in combinatorial geometry is not as great as one might suppose. The restraints in the former are the formation rules of English spelling and grammar, and those in the latter are the rules of mathematics. At least two eminent mathematicians, Augustus De Morgan and Claude E. Shannon, are on record as having spent considerable time composing pangrams, and I know of many lesser mathematicians who have tried their hand at it.

In mathematics a packing problem in general is one in which a given set of mathematical objects are to be packed as efficiently as possible into a given space according to given rules. Computer scientists, for example, are concerned with finding fast algorithms for packing sets of numbers into "bins," with the sum of the numbers in each bin not to exceed a specified limit. Such algorithms are needed for the efficient storage and retrieval of information. Geometrically the task can be viewed as a problem in one-dimensional packing: packing rods of varying lengths inside long pipes into which the rods fit snugly.

In a complex industrial society all kinds of problems arise involving the packing of three-dimensional objects into a specified area: the storing of objects in a warehouse; the packing of supplies into ships, planes, and freight cars; the packing of objects in cartons for distribution to stores, and so on. Perhaps it is the increasing need for packing algorithms that has stimulated some mathematicians in recent years to spend more time on such problems.

Here, we consider only the simplest kind of solid packing: the packing of "bricks" (rectangular parallelepipeds) into a "box" (also a rectangular parallelepiped). To simplify still more, we assume that all three dimensions of both bricks and box are integral, and that the volume of the box exactly equals the total volume of the bricks to be put inside. As David A. Klarner says in his article "Brick-packing Puzzles," many people are surprised to learn that even with these strong simplifications there are problems that are both elegant and challenging. By elegant Klarner means the following. If the bricks will pack the box perfectly, the problem is elegant if finding a way to pack it seems simple but is actually difficult. And if the bricks will not pack the box, the problem is elegant if there is a simple but subtle way to prove impossibility. Klarner, now at the University of Nebraska in Lincoln, is one of the pioneers of brick-packing theory. It is to him that I am indebted for most of what follows.

About 1960 the Dutch mathematician Nicolaas G. de Bruijn was struck by the fact that his son, age seven, was unable to fill a $6 \times 6 \times 6$ box with twenty-seven bricks, each $1 \times 2 \times 4$. The two volumes are the same and the packing seems easy, but one always ends up with at least one hole that the last brick will not fill. Studying the matter led de Bruijn to interesting results. They were first published as problems in a Hungarian journal, then later summarized by de Bruijn in his paper "Filling Boxes with Bricks".

De Bruijn calls a brick harmonic if its three measurements are integral and can be ordered so that each length is a multiple of the preceding one. In algebraic terms a harmonic brick has the form $a \times ab \times abc$, where the letters are positive integers. The $1 \times 2 \times 4$ brick is, of course, harmonic. It is called the canonical brick because it not only is the simplest harmonic brick of three distinct measurements but also approximates the shape of ordinary bricks used in masonry (*see* Figure 116).

De Bruijn was able to prove that a collection of identical harmonic bricks, each $a \times ab \times abc$, will perfectly pack a box if and only if the box is $ax \times aby \times abcz$. "Perfectly pack" means to fill completely; it is the same as saying that the brick will tile the box. De Bruijn showed that perfect packing is possible only if the box's dimensions are multiples of the brick's dimensions. To put it another way, if the bricks pack the box at all, there will be a way to do it trivially. That means they will pack when all are identically oriented. (Of course, they may also pack in nontrivial ways.) If the bricks are not harmonic, there are boxes they will fill only in a nontrivial way. For example, five nonharmonic bricks of $1 \times 2 \times 3$ will pack a $1 \times 5 \times 6$ box, but they cannot pack the box if they are all parallel.

De Bruijn's results generalize to hyperbricks in all higher Euclidean spaces, and they also hold for 2-space "bricks" (rectangles). The canonical plane brick—the 1×2 domino—will pack a rectangle only if one side is even, and then, of course, a trivial packing is possible.

Let us return to the task that puzzled de Bruijn's son. Since 6 is not a multiple of 4, we know from de Bruijn's work that the canonical brick will not pack an order-6 cubical box. Is there a simple impossibility proof?

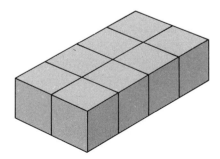

Figure 116 The canonical brick

There is, and it is a generalization of the solution to the old brainteaser about an order-8 checkerboard that has had two diagonally opposite corner squares removed. Can the board be covered with thirty-one dominoes? The fact that it cannot is evident once you realize that the two missing squares are the same color. The board therefore contains thirty-two squares of one color and thirty of the other. Since a domino must cover two squares of opposite color, after thirty dominoes are put down there will always be two uncovered squares of the same color that cannot be covered by the last domino.

To apply the same kind of parity check to the cube-packing problem, imagine that the order-6 cube is divided into twenty-seven order-2 cubes and that the order-2 cubes are shaded as is shown in Figure 117. No matter how a canonical brick is oriented inside such a cube, it will fill four shaded cells and four white cells. The cube, however, has eight more shaded cells than white

Figure 117 Color scheme for an order-6 cube-packing problem

cells. Therefore, after twenty-six bricks are placed, eight cells of the same color will remain. Clearly the last brick cannot fill them.

Will 125 canonical bricks pack a $10 \times 10 \times 10$ box? They will not, and the same impossibility proof applies. Indeed, the proof applies to any cube with a side that is even and not a multiple of 4. Will 250 bricks of $1 \times 1 \times 4$ pack the order-10 cube? As the impossibility proof shows, the answer is no. Sometimes more than two colors are needed for elegant impossibility proofs, but it is surprising how much can be done with just two colors.

Here is a delightful tiling problem from Klarner that is easily proved impossible by two-coloring, although not a checkerboard coloring. You have a 25×25 square you want to tile (no overlap, no vacancies) with a mixture of 2×2 squares and 3×3 squares. I shall give his impossibility proof in the answer section.

One of Klarner's theorems introduces the concept of cleavability. If a rectangle can be tiled with identical rectangles, there is always a way to tile it so that it can be cut into two smaller rectangles, each of which can also be tiled. Such a rectangle is said to be cleavable.

Does this unusual theorem have a 3-space analogue? That is, if a box can be fully packed with identical bricks, can it always be cut into two smaller boxes, each packable with the same bricks? The answer, Klarner found, surprisingly is no. The smallest example (discovered by David Singmaster) is the $5 \times 5 \times 12$ box. It can be packed with $1 \times 3 \times 4$ bricks, but not in a way that is cleavable.

Is there an infinite number of noncleavable boxes a given brick will pack? Again the answer, surprisingly, is no. Klarner was the first to show that in any Euclidean space an infinite set of boxes packable by a given brick has a finite subset of packable boxes that can be used for packing all the others. In addition, he showed that for every brick, there is a finite set of packable but noncleavable boxes. In 1971 two Hungarian mathematicians, G. Katona and D. Szász, refined Klarner's findings by giving a constructive proof that included specific numerical bounds.

If instead of identical bricks, we allow a mixture of different bricks (as in Klarner's tiling problem), many beautiful new problems arise. One of the simplest is a $3 \times 3 \times 3$ box puzzle that, Klarner says, first appeared in a Dutch book in 1970. We want to pack it with six $1 \times 2 \times 2$ bricks and three unit cubes (see Figure 118). It looks ridiculously easy, yet many find it irritatingly difficult. The reader is urged to construct a set, either by making the pieces out of wood or by gluing cubes together.

The packing has a unique solution (not counting rotations and reflections) requiring that the three unit cubes be along a space diagonal (see Figure 119).

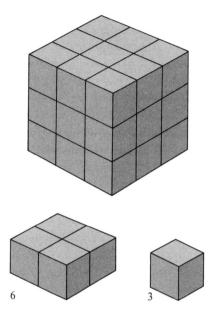

Figure 118   The 3 × 3 × 3 packing puzzle

To prove it, first consider any 3 × 3 cross section. If it is checkerboard-colored, five cells are of one color and four are of the other. No matter how a 1 × 2 × 2 brick is placed, it will occupy 0, 2, or 4 cells of each cross section, with the colors of the occupied cells evenly divided. As a result, each of the nine sections must have one and only one cell occupied by a unit cube. (A

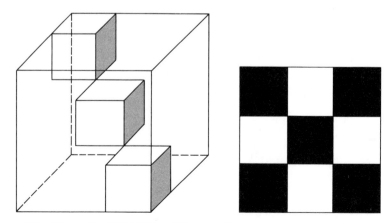

Figure 119   Key to the 3 × 3 × 3 packing puzzle

section cannot contain all three cubes because that would force some sections to be without cubes.) In addition, in every section the unit cube must be at the center or at one of the corners. The only way to meet these requirements is to place the cubes along a space diagonal. Placing the six bricks then follows automatically.

John Horton Conway of the University of Cambridge set himself the task a few years ago of designing a more difficult cube-packing puzzle. Conway's cube, although it appears to be almost as easy as the 3 × 3 × 3 puzzle, is so hard that some people cannot solve it until they are told its whimsical secret.

Conway's cube has many variants. The one he likes best requires eighteen harmonic bricks (*see* Figure 120). The task is to pack them into a 5 × 5 × 5 box or, what is the same thing, to build an order-5 cube. It is fiendishly difficult if one tries to solve it by trial and error.

Returning to the packing of identical bricks, we can ask the following general question. If a box is not perfectly packable with a set of bricks, what is the maximum number of bricks that will go into it? Even when the bricks are harmonic, it is an extremely difficult problem, although a good start has been made in a paper published in 1974 by Richard A. Brualdi and Thomas H. Foregger.

The authors define a "representing set," abbreviated R, as a set of cells in the box such that no matter where a brick is placed, it will occupy at least one cell of R. When R is made as small as possible, it is said to have minimum cardinality. The authors show that the maximum number of identical bricks

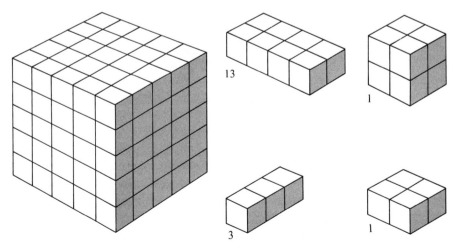

Figure 120 The eighteen bricks of John Horton Conway's cube-packing puzzle

(not necessarily harmonic) that will go into a box is equal to or less than the minimum cardinality of R.

In 2-space the maximum number of harmonic bricks (dominoes) invariably equals the minimum cardinality of R, but that does not hold for higher spaces. Consider canonical bricks and cubical boxes. The order-4 cube is the smallest that is packable, and of course the packing is trivial. The order-6 cube, as we have seen, is not packable, but all bricks but one are easily put inside.

What about the order-5 cube? It has 125 cells. That is not a multiple of 8, so that it is not fully packable by canonical bricks. Will fifteen such bricks (total volume 120) go inside? To put it another way, can you build an order-5 cube with fifteen canonical bricks and five unit cubes? The minimum cardinality of R is 15, but try as you will it is impossible to get fifteen bricks inside. After fourteen bricks have been placed, the thirteen remaining cells will never accommodate the last brick. Brualdi and Foregger have a complicated coloring proof of the impossibility, but in meditating about it one afternoon, I had a happy inspiration. It led to the following *reductio ad absurdum* proof.

Assume that fifteen bricks will pack. The total surface area of the order-5 cube is $6 \times 25 = 150$. Since each face has an odd number of squares, one cell on each face must be filled by a unit cube. (No more than one unit cube can occupy a cross section because there are fifteen such sections and only five unit cubes.) This leaves a surface of $150 - 6 = 144$ to be packed by faces of the canonical bricks. Now each brick must have one and only one of its 1×2 ends on the surface. This leaves a surface of $144 - (15 \times 2) = 114$ to be packed by 1×4 and 2×4 rectangles. But 114 is not evenly divisible by 4. Therefore the original assumption is false.

A similar proof is obtained by considering the cube's three planar midsections. Each plane is a 5×5 matrix, making seventy-five cells in all. One cell in each 5×5 square must be intersected by a unit cube. (This could be done with a single unit cube at the center, or two or three unit cubes suitably placed.) Each of the fifteen canonical bricks must intersect two cells, or thirty cells in all. Taking thirty-three cells from seventy-five leaves forty-two to be intersected in sets of four or eight, and since 42 is not a multiple of 4, impossibility follows.

The next cube of interest is the order 7. It is easy to put forty-one canonical bricks inside it, but will it hold forty-two, leaving seven holes to be filled by unit cubes? Surprisingly the answer is not known. Foregger posed this as an unsolved problem (E 2524) in the March 1975 issue of *American Mathematical Monthly*. Although every canonical brick must intersect a pair of cells in the planar midsections, after the necessary subtractions are made, the remaining cells are a multiple of 4. Therefore the previous impossibility proof does not apply.

Brualdi and Foregger found many special cases in which the maximum number of harmonic bricks equals the minimal cardinality of R. For example, if the smallest face of the brick packs each face of the box, there is equality. For canonical bricks there is equality if one of the box's dimensions is even. The general problem, however, is far from solved.

I can think of no better way to conclude than by quoting the final sentence of Klarner's article: "A word of warning in closing! Engaging in experiments with little wooden blocks is fraught with the danger that friends, family and colleagues will assume you are entering your second childhood, and that you should be put away. A good defense is to have a few copies of Conway's puzzle on hand to divert their attention while you make a getaway."

ANSWERS

Readers were asked to prove that it is impossible to tile a square of side 25 with a mixture of 2 × 2 and 3 × 3 squares. David A. Klarner's proof begins by two-coloring the square with stripes (*see* Figure 121). No matter how a 2 × 2 square is placed, it will cover two colored cells and two white ones. Since the large square contains an excess of twenty-five colored cells, the excess will remain no matter how many 2 × 2 squares are placed or where they are placed. If the square can be tiled, we must therefore find a way to place a set of 3 × 3 squares so that they cover twenty-five more colored cells than white ones.

It is apparent that a 3 × 3 square, however placed, covers either six colored cells and three white ones or six white cells and three colored ones. In each case, the difference is 3. But 25 is not a multiple of 3, and so there is no way the 3 × 3 squares can cover an excess of twenty-five colored cells. Therefore the tiling is impossible.

Klarner has shown in general that if a rectangle can be tiled at all by squares of a × a and b × b, it can always be split into two rectangles (one possibly empty), one of which is tilable by a × a squares alone and the other tilable by b × b squares alone.

The second task was to build a 5 × 5 × 5 cube with eighteen specified bricks. In working on puzzles of that type one tends to place large pieces first, then try to fit smaller ones into the gaps. In this case it is a fatal strategy. If you worked on the puzzle by trial and error, you probably found that if you left the three small bricks to the last, they never fitted. The puzzle is difficult precisely because those three pieces must be in a unique configuration, and it is unlikely that you would hit on it by chance.

Let us approach the problem by way of combinatorial geometry. If we checkerboard-color a 5 × 5 cross section, we find that it contains thirteen cells

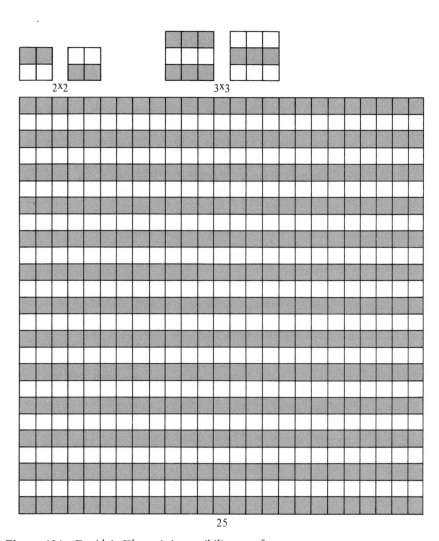

Figure 121 David A. Klarner's impossibility proof

of one color and twelve of the other (*see* Figure 122). Consider all the bricks except the three small ones. No matter how one brick is placed, it must fill an even number of cells (0, 2, 4, or 8) in every section. Half of the cells it fills will be one color and half will be the other.

Since there are an odd number of cells in each section, one or three of its cells must be occupied by a $1 \times 1 \times 3$ brick. In addition, the brick must be placed so that if it fills one cell, it will be a cell of the same color as the central cell. If it lies entirely within the section, two of its three cells must match the color of the central cell.

No matter how a small piece is placed, it will occupy five sections. It follows that the three pieces, if they contribute to all fifteen sections (as they must), will have to be placed so that no two contribute to the same section. There is only one way to do that and at the same time meet the coloring requirements. Once the small pieces are correctly positioned, it is not hard to find one of the more than 500 ways of packing the larger bricks around them.

John Horton Conway, who invented this cube puzzle, could have substituted another $1 \times 2 \times 4$ for the $2 \times 2 \times 2$, thus providing the maximum number of canonical bricks that will go into a $5 \times 5 \times 5$ box. Adding the order-2 cube, however, provides amusing misdirection. The puzzle can also be given with twenty-nine bricks of $1 \times 2 \times 2$, together with the three $1 \times 1 \times 3$ bricks. Whether most people would find that easier or more difficult is hard to say. The principle behind Conway's cube generalizes in the sense that three $1 \times 1 \times (2n - 1)$ bricks have a unique configuration in a cubical box of side $2n + 1$ that makes it possible to pack the rest of the box with $1 \times 2 \times 2$ bricks.

ADDENDUM

The chapter posed an unsolved problem about the packing of "canonical bricks" ($1 \times 2 \times 4$) into cubes of side n, with n greater than 3. When n is even, the cube can be fully packed (in a trivial way) only if n is a multiple of 4. I gave a simple proof that if n is even and not a multiple of 4, complete packing is impossible. It is necessary to omit one brick.

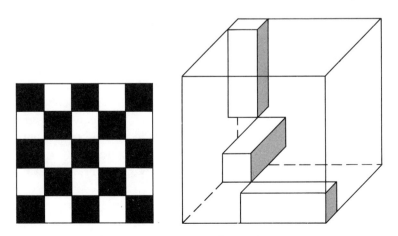

Figure 122 Key to John Horton Conway's cube puzzle

When n is odd, the situation is more interesting. Each $n \times n$ "slab" must contain a unit hole; consequently the maximum number of bricks that an odd-order cube can take is $(n^3 - n)/8$. For what values of n can this maximum packing be achieved?

When I wrote, only the 5-cube had been proved impossible. The proof was later generalized by David A. Klarner to cover exactly half of all odd n, namely cubes with sides equal to plus or minus 1 (modulo 8). Many impossibility proofs for the 7-cube were found by readers, but they did not generalize.

The problem is now solved. Robert Ammann of Lowell, Mass., first proved impossibility of maximal packing for all odd n except n equal to plus or minus 1 (modulo 16). He then found a maximal packing for the 15-cube that generalized to provide maximal packings for all odd n equal to plus or minus 1 (modulo 16).

Frank Barnes, working independently in England, obtained Ammann's earlier impossibility result and conjectured that maximum packings exist for the 15-cube and all others with sides plus or minus 1 (modulo 16). Told of Ammann's work, Barnes verified its soundness. In addition, he found a way to show that all impossible cubes *could* be packed by omitting just one brick from the maximum number. It is hoped that all these results will eventually be published.

Neither Ammann nor Barnes has an academic post. Ammann is a computer programmer working in nonmathematical areas. Although Barnes has lectured on mathematics at the University of Michigan and the University of Reading, he has been earning a living for the past two years by piloting a hot-air balloon for advertising purposes.

After the above comments ran in my column for October 1976, a proof by Michael Mather, that the order-7 box could not be packed with forty-two canonical bricks, was published (*The American Mathematical Monthly* 83, November 1976, pp. 741–742) as a solution to Problem E 2524. Mather showed more generally that it is possible to pack a cube of side $2n + 1$ with $2n(2n + 1)$ bricks, each of size $1 \times 2 \times (n + 1)$, if and only if n is even or equal to 1.

David Klarner found what he calls a "divide and conquer" way to put forty-one canonical bricks into the order-7 cube. Cut the cube into three sub-boxes $2 \times 7 \times 7$, $3 \times 5 \times 7$, and $4 \times 5 \times 7$. "It is easy," he writes in a 1987 letter, "to pack twelve bricks into the $2 \times 7 \times 7$, and even easier to pack seventeen into the $4 \times 5 \times 7$." Putting twelve bricks into the $3 \times 5 \times 7$ is not so easy, but it can be done. There are nine holes. One arrangement has three $1 \times 1 \times 3$ holes.

Klarner uses the term "atomic" for a box (not necessarily a cube) that cannot be optimally packed by divide and conquer. In other words, no maximal packing will split into smaller boxes that can be optimally packed. The most interesting packing problems involve atomic boxes. We can ask: does the canonical brick have an infinite or finite number of atomic boxes it will pack?

The question is open. Some years ago Klarner proved that in the case of *tiling* a box (packing completely with no holes), for every brick there is a finite set of boxes with integer sides that are atomic. For several years he thought a similar theorem applied to optical packing with holes. That is, for any given brick there is a finite number of atomic boxes it will optimally pack.

"How wrong I was!" he writes. He has since proved that for the $1 \times 2 \times 2$ brick there is an infinite number of atomic cubical boxes it will pack optimally.

Much remains unknown about packing canonical bricks into noncubical boxes. Klarner tells me that his student Wade Satterfield has shown that twenty-one bricks cannot fit into a $5 \times 5 \times 7$ box. (It is a simple, clever proof, not yet published, based on a parity contradiction.) The same proof also shows that twenty-seven canonical bricks will not pack the $5 \times 5 \times 9$. Will the proof generalize to show that $3n$ such bricks will not pack a $5 \times 5 \times n$ box? No, writes Klarner, it breaks down for all odd n greater than 9.

"It is an open question," he continues, "whether there exists any odd number n such that one can pack $3n$ canonical bricks into a $5 \times 5 \times n$ box. Since one can pack six canonical bricks into a $5 \times 5 \times 2$ box, it follows that if one can pack $3k$ such bricks into a $5 \times 5 \times k$ box for some odd number k, then this can be done for all numbers larger than k."

B I B L I O G R A P H Y

"Filling Boxes with Bricks." N. G. de Bruijn, in *The American Mathematical Monthly* 76, January 1969, pp. 37–40.

"Matching Problems." G. Katona and D. Szász, in *Journal of Combinatorial Theory* 10, February 1971, pp. 60–92.

"Brick-Packing Puzzles." David A. Klarner, in *Journal of Recreational Mathematics* 6, Spring 1973, pp. 112–117.

"Packing Boxes with Harmonic Bricks." Richard A. Brualdi and Thomas H. Foregger, in *Journal of Combinatorial Theory* 17, October 1974, pp. 81–114.

"A Finite Basis Theorem for Packing Boxes with Bricks." N. G. de Bruijn and David A. Klarner, in *Philips Research Reports* 30, 1975, pp. 337–343.

"Box Packing Problems." Ross Honsberger, in *Mathematical Gems 2,* Chapter 8. Mathematical Association of America, 1976.

"Packing Problems and Inequalities." D. G. Hoffman, in *The Mathematical Gardner.* David A. Klarner, ed. Prindle, Weber and Schmidt, 1981.

NINETEEN

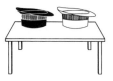

.
.
.

Induction and Probability

"The universe, so far as known to us,
is so constituted, that whatever is
true in any one case, is true in all
cases of a certain description; the only
difficulty is, to find what description."

—JOHN STUART MILL,
A System of Logic

Imagine that we are living on an intricately patterned carpet. It may or may
not extend to infinity in all directions. Some parts of the pattern appear to be
random, like an abstract expressionist painting; other parts are rigidly geometrical. A portion of the carpet may seem totally irregular, but when the same
portion is viewed in a larger context, it becomes part of a subtle symmetry.

The task of describing the pattern is made difficult by the fact that the carpet
is protected by a thick plastic sheet with a translucence that varies from place to
place. In certain places we can see through the sheet and perceive the pattern;

in others the sheet is opaque. The plastic sheet also varies in hardness. Here and there we can scrape it down so that the pattern is more clearly visible. In other places the sheet resists all efforts to make it less opaque. Light passing through the sheet is often refracted in bizarre ways, so that as more of the sheet is removed, the pattern is radically transformed. Everywhere there is a mysterious mixing of order and disorder. Faint lattices with beautiful symmetries appear to cover the entire rug, but how far they extend is anyone's guess. No one knows how thick the plastic sheet is. At no place has anyone scraped deep enough to reach the carpet's surface, if there is one.

Already the metaphor has been pushed too far. For one thing, the patterns of the real world, as distinct from this imaginary one, are constantly changing, like a carpet that is rolling up at one end while it is unraveling at the other end. Nevertheless, in a crude way the carpet can introduce some of the difficulties philosophers of science encounter in trying to understand why science works.

Induction is the procedure by which carpetologists, after examining parts of the carpet, try to guess what the unexamined parts look like. Suppose the carpet is covered with billions of tiny triangles. Whenever a blue triangle is found, it has a small red dot in one corner. After finding thousands of blue triangles, all with red dots, the carpetologists conjecture that all blue triangles have red dots. Each new blue triangle with a red dot is a confirming instance of the law. Provided that no counterexample is found, the more confirming instances there are, the stronger is the carpetologists' belief that the law is true.

The leap from "some" blue triangles to "all" is, of course, a logical fallacy. There is no way to be absolutely certain, as one can be in working inside a deductive system, what any unexamined portion of the carpet looks like. On the other hand, induction obviously works, and philosophers justify it in other ways. John Stuart Mill did so by positing, in effect, that the carpet's pattern has regularities. He knew this reasoning was circular, since it is only by induction that carpetologists have learned that the carpet is patterned. Mill did not regard the circle as vicious, however, and many contemporary philosophers (R. B. Braithwaite and Max Black, to name two) agree. Bertrand Russell, in his last major work, tried to replace Mill's vague "nature is uniform" with something more precise. He proposed a set of five posits about the structure of the world that he believed were sufficient to justify induction.

Hans Reichenbach advanced the most familiar of several pragmatic justifications. If there is any way to guess what unexamined parts of the carpet look like, Reichenbach argued, it has to be by induction. If induction does not work, nothing else will, and so science might as well use the only tool it has. "This answer is not fallacious," wrote Russell, "but I cannot say that I find it very satisfying."

Rudolf Carnap agreed. His opinion was that all these ways of justifying induction are correct but trivial. If "justify" is meant in the sense that a mathematical theorem is justified, then David Hume was right: There is no justification. But if "justify" is taken in any of several weaker senses, then, of course, induction can be defended. A more interesting task, Carnap insisted, is to see whether it is possible to construct an inductive logic.

It was Carnap's great hope that such a logic could be constructed. He foresaw a future in which a scientist could express in a formalized language a certain hypothesis together with all the relevant evidence. Then by applying inductive logic, he could assign a probability value, called the degree of confirmation, to the hypothesis. There would be nothing final about that value. It would go up or down or stay the same as new evidence accumulated. Scientists already think in terms of such a logic, Carnap maintained, but only in a vague, informal way. As the tools of science become more powerful, however, and as our knowledge of probability becomes more precise, perhaps eventually we can create a calculus of induction that will be of practical value in the endless search for scientific laws.

In Carnap's *Logical Foundations of Probability* (University of Chicago Press, 1950) and also in his later writings, he tried to establish a base for such a logic. How successful he was is a matter of dispute. Some philosophers of science share his vision (John G. Kemeny for one) and have taken up the task where Carnap left off. Others, notably Karl Popper and Thomas S. Kuhn, regard the entire project as having been misconceived.

Carl G. Hempel, one of Carnap's admirers, has argued sensibly that before we try to assign quantitative values to confirmations, we should first make sure we know in a qualitative way what is meant by "confirming instance." It is here that we run into the worst kinds of difficulty.

Consider Hempel's notorious paradox of the raven. Let us approach it by way of 100 playing cards. Some of them have a picture of a raven drawn on the back. The hypothesis is: "All raven cards are black." You shuffle the deck and

A cartoon comment on inductive reasoning

deal the cards face up. After turning fifty cards without finding a counterinstance, the hypothesis certainly becomes plausible. As more and more raven cards prove to be black, the degree of confirmation approaches certainty and may finally reach it.

Now consider another way of stating the same hypothesis: "All nonblack cards are not ravens." This statement is logically equivalent to the original one. If you test the new statement on another shuffled deck of 100 cards, holding them face up and turning them as you deal, clearly each time you deal a nonblack card and it proves to have no raven on the back, you confirm the guess that all nonblack cards are not ravens. Since this is logically equivalent to "All raven cards are black," you confirm that also. Indeed, if you deal all the cards without finding a red card with a raven, you will have completely confirmed the hypothesis that all raven cards are black.

Unfortunately, when this procedure is applied to the real world, it seems not to work. "All ravens are black" is logically the same as "All nonblack objects are not ravens." We look around and see a yellow object. Is it a raven? No, it is a buttercup. The flower surely confirms (albeit weakly) that all nonblack objects are not ravens, but it is hard to see how it has much relevance to "All ravens are black." If it does, it also confirms that all ravens are white or any color except yellow. To make things worse, "All ravens are black" is logically equivalent to "Any object is either black or not a raven." And that is confirmed by any black object whatever (raven or not) as well as by any nonraven (black or not). All of which seems absurd.

Nelson Goodman's "grue" paradox is equally notorious. An object is "grue" if it is green until, say, January 1, 2000, and blue thereafter. Is the law "All emeralds are grue" confirmed by observations of green emeralds? A prophet announces that the world will exist until January 1, 2000, when it will disappear with a bang. Every day the world lasts seems to confirm the prediction, yet no one supposes that it becomes more probable.

To make matters still worse, there are situations in which confirmations make a hypothesis less likely. Suppose you turn the cards of a shuffled deck looking for confirmations of the guess that no card has green pips. The first ten cards are ordinary playing cards, then suddenly you find a card with blue pips. It is the eleventh confirming instance, but now your confidence in the guess is severely shaken. Paul Berent has pointed out several similar examples. A man 99 feet tall is discovered. He is a confirming instance of "All men are less than 100 feet tall," yet his discovery greatly *weakens* the hypothesis. Finding a normal-size man in an unlikely place (such as Saturn's moon Titan) is another example of a confirming instance that would weaken the same hypothesis.

Confirmations may even falsify a hypothesis. Ten cards with all values from the ace through the 10 are shuffled and dealt face down in a row. The guess is that no card with value n is in the nth position from the left. You turn the first nine cards. Each card confirms the hypothesis. But if none of the turned cards is the 10, the nine cards taken together refute the hypothesis.

Here is another example. Two piles of three cards each are on the table. One pile consists of the jack, queen, and king of hearts, the other of the jack, queen, and king of clubs. Each has been shuffled. Smith draws a card from the heart pile, Jones takes a card from the club pile. The hypothesis is that the pair of selected cards consists of a king and queen. The probability of this is 2/9. Smith looks at his card and sees that it is a king. Without naming it, he announces that his card has confirmed the hypothesis. Why? Because knowing that his card is a king raises the probability of the hypothesis being true from 2/9 to 3/9 or 1/3. Jones now sees that he (Jones) has drawn a king, so he can make the same statement Smith made. Each card, taken in isolation, is a confirming instance. Yet, both cards taken together falsify the hypothesis.

Carnap was aware of such difficulties. He distinguished sharply between "degree of confirmation," a probability value based on the total relevant evidence, and what he called "relevance confirmation," which has to do with how new observations alter a confirmation estimate. Relevance confirmation cannot be given simple probability values. It is enormously complex, swarming with counterintuitive arguments. In Chapter 6 of Carnap's *Logical Foundations* he analyzes a group of closely related paradoxes of confirmation relevance that are easily modeled with cards.

For example, it is possible that data will confirm each of two hypotheses but disconfirm the two taken together. Consider a set of ten cards, half with blue backs and half with green ones. The green-backed cards (with the hearts and spades designated H and S) are QH, $10H$, $9H$, KS, QS. The blue-backed cards are KH, JH, $10S$, $9S$, $8S$. The ten cards are shuffled and dealt face down in a row.

Hypothesis A is that the property of being a face card (a king, a queen, or a jack) is more strongly associated with green backs than with blue. An investigation shows that this is true. Of the five cards with green backs, three are face cards as against only two face cards with blue backs. Hypothesis B is that the property of being a red card (hearts or diamonds) is also more strongly associated with green backs than with blue. A second investigation confirms this. Three green-backed cards are red, but there are only two red cards with blue backs. Intuitively one assumes that the property of being both red and a face card is more strongly associated with green backs than blue, but that is not the case. Only one red face card has a green back, whereas two red face cards have blue backs!

It is easy to think of ways, fanciful or realistic, in which similar situations can arise. A woman wants to marry a man who is both rich and kind. Some of the bachelors she knows have hair and some are bald. Being a statistician, she does some sampling. Project *A* establishes that 3/5 of the men with hair are rich but only 2/5 of the bald men are rich. Project *B* discloses that 3/5 of the men with hair are kind but only 2/5 of the bald men are kind. The woman might hastily conclude that she should marry a man with hair, but if the distribution of the attributes corresponds to that of the face cards and red cards mentioned in the preceding example, her chances of getting a rich, kind man are twice as great if she sets her cap for a bald man.

Another research project shows that 3/5 of a group of patients taking a certain pill are immune to colds for five years, compared with only 2/5 in the control group who were given a placebo. A second project shows that 3/5 of a group receiving the pill were immune to tooth cavities for five years, compared with 2/5 who got the placebo. The combined statistics could show that twice as many among those who got the placebo are free for five years from both colds and cavities, compared with those who got the pill.

A striking instance of how a hypothesis can be confirmed by two independent studies, yet disconfirmed by the total results, is provided by the following game. It can be modeled with cards, but to vary the equipment let us do it with forty-one poker chips and four hats (*see* Figure 123). On table *A* is a black hat containing five colored chips and six white chips. Beside it is a gray hat containing three colored chips and four white chips. On table *B* is another pair of black and gray hats. In the black hat there are six colored chips and three white chips. In the gray hat there are nine colored chips and five white chips. The contents of the four hats are shown by the charts in the illustration.

You approach table *A* with the desire to draw a colored chip. Should you take a chip from the black hat or from the gray one? In the black hat five of the eleven chips are colored, so that the probability of getting a colored chip is 5/11. This is greater than 3/7, which is the probability of getting a colored chip if you take a chip from the gray hat. Clearly your best bet is to take a chip from the black hat.

The black hat is also your best choice on table *B*. Six of its nine chips are colored, giving a probability of 6/9, or 2/3, that you will get a colored chip. This exceeds the probability of 9/14 that you will get a colored chip if you choose to take a chip from the gray hat.

Now suppose that the chips from both black hats are combined in one black hat and that the same is done for the chips in the two gray hats (*see* Table *C* in Figure 123.) If you want to get a colored chip, surely you should take a chip from the black hat. The astonishing fact is that this is not true! Of the twenty chips now in the black hat, eleven are colored, giving a probability of 11/20 that you will get a colored chip. This is exceeded by a probability of 12/21 that you will get a colored chip if you take a chip from the gray hat.

The situation has been called Simpson's paradox by Colin R. Blyth, who found it in a 1951 paper by E. H. Simpson. The paradox has turned out to be older, but the name has persisted. Again, it is easy to see how the paradox could arise in actual research. Two independent investigations of a drug, for example, might suggest that it is more effective on men than it is on women, whereas the combined data would indicate the reverse.

One might imagine that such situations are too artificial to arise in statistical research. In a recent investigation to see if there was sex bias in the admissions of men and women to graduate studies at the University of California at Berkeley, however, Simpson's paradox actually turned up. (*See* "Sex Bias in Graduate Admissions: Data from Berkeley," by P. J. Bickel, E. A. Hammel, and J. W. O'Connell.)

Blyth has invented another paradox that is even harder to believe than Simpson's. It can be modeled with three sets of cards or three unfair dice that are weighted to give the required probability distributions to their faces. We shall model it with the three spinners shown in Figure 124, because they are easy to construct by anyone who wants to verify the paradox empirically.

Spinner *A,* with an undivided dial, is the simplest. No matter where the arrow stops, it gives a value of 3. Spinner *B* gives values of 2, 4, or 6 with the respective probability distributions of .56, .22, and .22. Spinner *C* gives values of 1 or 5 with the probabilities of .51 and .49.

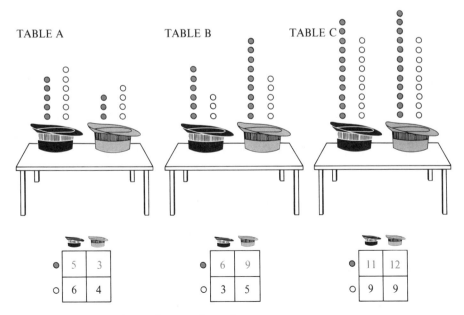

Figure 123   E. H. Simpson's reversal paradox

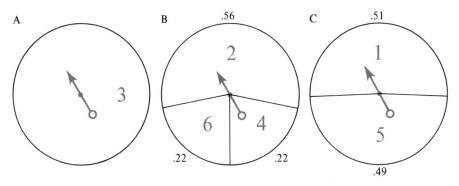

Figure 124 The three spinners for Colin R. Blyth's paradox

You pick a spinner; a friend picks another. Each of you flicks his arrow, and the highest number wins. If you can later change spinners on the basis of experience, which spinner should you choose? When the spins are compared in pairs, we find that A beats B with a probability of $1 \times .56 = .56$. A beats C with a probability of $1 \times .51 = .51$. B beats C with a probability of $(1 \times .22) + (.22 \times .51) + (.56 \times .51) = .6178$. Clearly A, which beats both of the others with a probability of more than $1/2$, is the best choice. C is the worst because it is beaten with a probability of more than $1/2$ by both of the others.

Now for the crunch. Suppose you play the game with two others and you have the first choice. The three spinners are flicked, and the high number wins. Calculating the probabilities reveals an extraordinary fact. A is the worst choice; C is the best! A wins with a probability of $.56 \times .51 = .2856$, or less than $1/3$. B wins with a probability of $(.44 \times .51) + (.22 \times .49) = .3322$, or almost $1/3$. C wins with a probability of $.49 \times .78 = .3822$, or more than $1/3$.

Consider the havoc this can wreak in statistical testing. Assume that drugs for a certain illness are rated in effectiveness with numbers 1 through 6. Drug A is uniformly effective at a value of 3 (spinner A). Studies show that drug C varies in effectiveness. Fifty-one percent of the time it has value 1, and 49 percent of the time it has value 5 (spinner C). If drugs A and C are the only two on the market and a doctor wants to maximize a patient's chance of recovery, he clearly chooses drug A.

What happens when drug B, with values and a probability distribution corresponding to spinner B, becomes available? The bewildered doctor, if he considers all three drugs, finds C preferable to A.

Blyth has an even more mind-blowing way of dramatizing the paradox. Every night, a statistician eats at a restaurant that offers apple pie and cherry pie. He rates his satisfaction with each kind of pie in values 1 through 6. The apple pie is uniformly 3 (spinner A); the cherry varies in the manner of spinner C. Naturally, the statistician always takes apple.

Occasionally the restaurant has blueberry pie. Its satisfaction varies in the manner of spinner B.

Waitress: Shall I bring your apple pie?

Statistician: No. Seeing that today you also have blueberry, I'll take the cherry.

The waitress would consider that a joke. Actually, the statistician is rationally maximizing his expectation of satisfaction. (An error. *See* Addendum.) Is there any paradox that points up more spectacularly the kinds of difficulty Carnap's followers must overcome in their efforts to advance his program?

ADDENDUM

Many readers quite properly chided me for carelessness when I described Colin R. Blyth's paradox of the man and the three pies. It was I (not Blyth) who said that the man's decision was to maximize his "expectation of satisfaction." What he is maximizing is, in Blyth's words, "his best chance" of getting the most satisfying pie. It is a subtle but important difference. Both the dining statistician and the doctor have a choice between two intents: maximizing their average of satisfaction in the long run or maximizing their chance of getting the best pie or drug on a particular occasion.

To put it another way, Blyth's pie eater is minimizing his regret: the probability that he will see a better pie on the next table. His doctor counterpart, as Paul Chernick suggested, could be trying to avoid a malpractice suit that might result if a dissatisfied patient went to another doctor and got more effective treatment. "Is the case of a scientist closer to that of a player in the spinner game," asked George Mavrodes, "or is it closer to that of the statistical pie eater? . . . I do not know the answer to that question."

John F. Hamilton, Jr., revised the dialogue between the waitress and the statistician as follows:

Waitress: Which pie will be better tonight, A or B?

Statistician: The odds are on A.

Waitress: What about A and C?

Statistician: Again, *A* will probably win.

Waitress: I see, you mean *A* will probably be the best of all.

Statistician: No, actually *C* has the greatest chance of being the best.

Waitress: Okay, cut the funny stuff. Which pie do you want to order, *A* or *C*?

Statistician: Neither. I'll have a slice of *B*, please.

The paradoxes of confirmation are not, of course, paradoxes in the sense of being contradictions, but paradoxes in the wider sense of being counterintuitive results that make nonsense of earlier naïve attempts, by John Stuart Mill and others, to define the meaning of "confirming instance." Philosophers who discuss the paradoxes are not ignorant of statistical theory. It is precisely because statistical theory demands so many careful distinctions that the task of formulating an inductive logic is so difficult.

Richard C. Jeffrey, in *The Logic of Decision Making* (University of Chicago Press, 2d ed. 1983), formulates an amusing variant of Goodman's "grue" paradox. We define a "goy" as a girl born before 2000 or a boy born after that date, and a "birl" as a boy born before 2000 or a girl born after that date. Until now, no goy has had a penis, and all birls have. Hence by induction, (*A*) the first goy born after 2000 will have no penis, and (*B*) the first birl born after that date will. However, the first goy born after 2000 will be a boy, which contradicts *A*. Similarly, the first birl born after 2000 will be a girl, which contradicts *B*.

B I B L I O G R A P H Y

Confirmation Theory

"Disconfirmation by Positive Instances." Paul Berent, in *Philosophy of Science* 39, December 1972, p. 522.

An Introduction to Confirmation Theory. Richard Swinburne. Methuen, 1973.

"Confirmation." Wesley C. Salmon, in *Scientific American,* May 1973, pp. 75–83.

Confirmation and Confirmability. George Schlesinger. Oxford University Press, 1974.

Simpson's Paradox

"On Simpson's Paradox and the Sure-Thing Principle." Colin R. Blyth, in *Journal of the American Statistical Association* 67, June 1972, pp. 364–381.

"Baseball Statistics." Edwin F. Beckenbach, in *Mathematics Teacher* 72, May 1979, pp. 351–352.

"Magic Possibilities of the Weighted Average." Ruma Falk and Maya Bar-Hillel, in *Mathematics Magazine* 53, March 1980, pp. 106–107.

"Simpson's Paradox in Real Life." Clifford Wagner, in *American Statistician* 36, February 1982, pp. 46–48.

"Sex Bias in Graduate Admissions: Data from Berkeley." P. J. Bickel, E. A. Hammel, and J. W. O'Connell, in *Science* 187, February 1985, pp. 398–404.

"Instances of Simpson's Paradox." Thomas R. Knapp, in *The College Mathematics Journal* 16, June 1985, pp. 209–211.

TWENTY

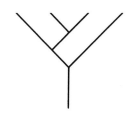

Catalan Numbers

If an infinite sequence of positive integers is simple enough, such as the doubling series (1, 2, 4, 8, 16, . . .) or the squares (1, 4, 9, 16, 25, . . .), it is easily recognized. And few mathematicians would fail to recognize the Fibonacci numbers (1, 1, 2, 3, 5, 8, . . .) or the triangular numbers (1, 3, 6, 10, 15, 21, . . .). If the sequence is unfamiliar, however, an enormous amount of time can be wasted searching for a recursive or nonrecursive procedure that generates the sequence. (A procedure is recursive if calculating a next term calls for knowledge of the preceding terms; a nonrecursive formula gives the nth term without such knowledge.)

It is hard to believe, but it was not until 1973 that *A Handbook of Integer Sequences* (Academic Press, 1973) was published. This invaluable tool, compiled by N. J. A. Sloane of Bell Laboratories, lists more than 2300 integer sequences in numerical order. A mathematician who encounters a puzzling sequence no longer needs to spend hours trying to find its generating formula. He simply looks for the sequence in Sloane's book. The chances are excellent that it is there, followed by a list of references where the reader can check on the nature of the beast.

Our topic here is the *Handbook's* sequence 577: 1, 2, 5, 14, 42, 132, 429, 1430, 4862, 16796, ... The components of this sequence are called Catalan numbers. They are not as well known as Fibonacci numbers, but they have the same delightful propensity for popping up unexpectedly, particularly in combinatorial problems. In 1971 Henry W. Gould, a mathematician at West Virginia University, privately issued a bibliography of 243 references on Catalan numbers; in many cases the authors were not even aware they were dealing with a sequence known for more than two centuries. In 1976 Gould increased the number of references to 450. Indeed, the Catalan sequence is probably the most frequently encountered sequence that is still obscure enough to cause mathematicians, lacking access to Sloane's *Handbook,* to expend inordinate amounts of energy rediscovering formulas that were worked out long ago.

It was Leonhard Euler who first discovered the Catalan numbers after asking himself: In how many ways can a fixed convex polygon be divided into triangles by drawing diagonals that do not intersect? An example can be provided with triangles, quadrilaterals, pentagons, and hexagons (*see* Figure 125). Note that in every case, regardless of how the n-gon is triangulated, the number of diagonals is always $n - 3$ and the number of triangles is $n - 2$. It is easy to prove that this relation holds in general. The number of possible triangulations for each of these four polygons are the first four terms of the Catalan sequence.

Applying an induction process that he described as "quite laborious," Euler obtained the following recursive formula:

$$\frac{2 \times 6 \times 10 \times \cdots (4n - 10)}{(n - 1)!}$$

Numbers above the line have the form $(4n - 10)$, where n is a positive integer greater than 2. The exclamation mark is of course the factorial sign. It stands for the product of all positive integers from 1 through the preceding expression. For example, if $n = 6$ (the sides of a hexagon), the formula becomes

$$\frac{2 \times 6 \times 10 \times 14}{5!} = 14$$

Unusually simple recursive formulas are obtained by putting another 1 in front of the series: 1, 1, 2, 5, 14, ... Let k be the last number of a partial sequence and n the position of the next number. The next number is then

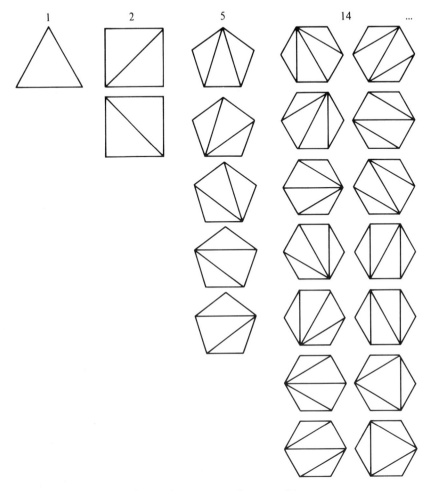

Figure 125 Leonhard Euler's polygon-triangulation problem

$$\frac{k(4n - 6)}{n}$$

Johann Andreas von Segner, Euler's eighteenth-century contemporary, found a whimsical recursive procedure for the same form of the Catalan sequence. Write the partial sequence forward, then put below it the same numbers in backward order. Multiply each top number by the one below it and add all the products; the result is the next number of the sequence. For example,

$$\begin{array}{ccccc} 1 & 1 & 2 & 5 & 14 \\ \times 14 & 5 & 2 & 1 & 1 \\ \hline \end{array}$$
$$14 + 5 + 4 + 5 + 14 = 4\overset{\bullet}{2}$$

Euler's polygon triangulation is isomorphic with many seemingly unrelated problems. It was Eugene Charles Catalan, the Belgian mathematician for whom the sequence is named, who in 1838 solved the following problem. We have a chain of n letters in a fixed order. We want to add $n - 1$ pairs of parentheses so that inside each pair of left and right parentheses there are two "terms." These paired terms can be any two adjacent letters, or a letter and an adjacent parenthetical grouping, or adjacent groupings. In how many ways can the chain be parenthesized?

For two letters, ab, there is only one way: (ab). For three letters there are two ways: $((ab)c)$ and $(a(bc))$. For four letters there are five ways: $((ab)(cd))$, $(((ab)c)d)$, $(a(b(cd)))$, $(a((bc)d))$, and $((a(bc))d)$. The numbers of these ways, 1, 2, and 5, are the first three Catalans, and the Catalan sequence enumerates the ways of parenthesizing all longer chains.

H. G. Forder, writing on Catalan numbers in 1961, showed a simple way to establish one-to-one correspondence between the triangulated polygons and the parenthesized expressions. An example is a triangulated heptagon (*see* Figure 126). Label its sides (excluding the base) a through f. Every diagonal spanning adjacent sides is labeled with the letters of those sides in parentheses. Each remaining diagonal is then lettered in similar fashion by combining the labels on the other two sides of the triangle. The base is lettered last. The

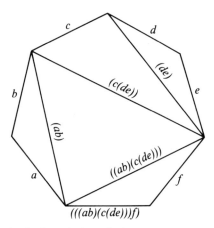

Figure 126 Parenthesized triangulation of a heptagon

expression for the base is uniquely determined by the dissection. If you apply this technique to the polygons portrayed in Figure 125, you will obtain the parenthesized expressions shown at the right in Figure 127.

The British mathematician Arthur Cayley proved that Catalan numbers count the number of trees that are planar, trivalent, and planted. A "tree" is a connected graph (points joined by edges) that has no circuits. "Planar" means that it can be drawn on the plane without intersections. "Planted" means that it has one "trunk," the end of which is called the "root." The graph can thus be drawn to simulate a tree growing up from the ground. "Trivalent" means that at each point (except at the root and at the ends of branches), the tree forks to create a spot where three edges meet.

The illustration of this structure (Figure 127) is almost self-explanatory. The gray lines show how each triangulation corresponds to a planted trivalent tree. Next to the polygons, corresponding trees are drawn in conventional form. It is easy to see how the grouping of a tree's branches corresponds to its parenthesized expression. Below each expression, we convert it to a binary number by replacing every left-hand parenthesis with 1 and every letter with zero, ignoring all right-hand parentheses. These binary numbers are convenient shorthand ways of designating the polygon dissections and their trees. Right-hand parentheses are not needed because, given the left-hand ones and the method of grouping letters, the right-hand parentheses can always be added in a unique manner.

The Polish mathematician Jan Lukasiewicz found a pleasant way to obtain each tree's binary number (*see* Figure 128). Picture a tree with four top ends. They are labeled 0 and the trivalent points are labeled 1. Imagine a worm crawling up the trunk and around the entire tree along the broken path in the illustration. At each point, the worm calls out the label. Once a point is called, it is not called again. In this example the worm calls out 1101000, which proves to be the very binary number we obtained from the tree's parenthesized expression.

In 1964 it was discovered that normal planted trees are also counted by Catalan numbers. They are planted trees of n points, including the ends but not the root. They can also be described as planted trees of n edges. A point in such a tree can have any valence.

Many ways have been found for showing a one-to-one correspondence between trees of this kind and the planted trivalent trees. The simplest one was pointed out by Frank Bernhart (*see* Figure 129). The trivalent trees are drawn so that at each point of valence 3 the edges go up, down, and to the right. Imagine that each horizontal edge shrinks to a point and disappears. If there is a trivalent point at the right end of the edge, it is carried to the left to merge

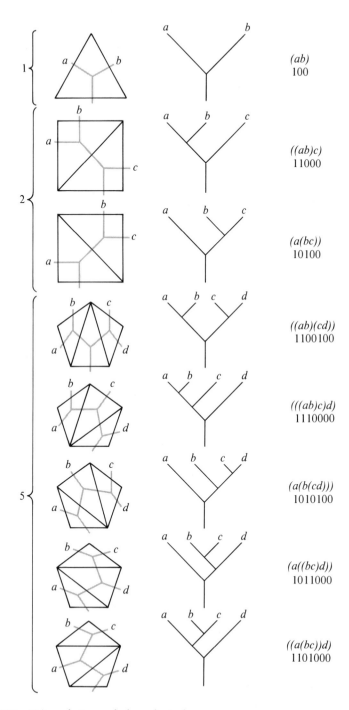

Figure 127 Triangulation and planted trivalent trees

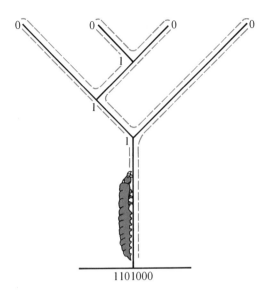

Figure 128 A worm generating a tree's binary number

with the point at the left. All the vertical edges remain distinct. This simple transformation changes all planted trivalent trees of *n* ends into all planted trees of *n* edges.

A worm crawling up and around any tree in this new set (the trees at the right in the illustration) will call out the same binary number as the tree's partner if it alters its procedure as follows. The worm starts at the bottom point rather than at the root. Each time it crawls up an edge, it calls 1, and each time it crawls down an edge, it calls 0.

Consider chessboards of sides 2, 3, 4, ... All squares north and west of a main diagonal are shaded (*see* Figure 130). We are to move the rook from the lower left corner to the upper right corner. It cannot enter a shaded cell, and its only allowed movements are north or east. For a board of side *n*, how many different paths can the rook take?

Once more the Catalans give the answer. Below each board of side *n* write the binary number for the planted trivalent tree of *n* ends. Taking the binary digits from left to right, move the rook one square to the right for each 1 and one square up for each 0. (The final digit is ignored.) This pattern generates a path, and in this way all the rook paths are obtained.

Here are seven more recreational problems solved by the Catalans. For the first five, I shall indicate how the corresponding binary numbers (ignoring final digits) solve the problem.

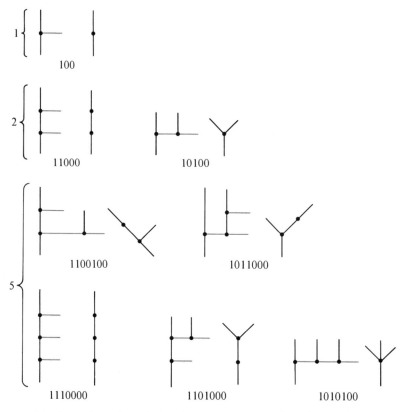

Figure 129 Transformation of planted trivalent trees to planted normal trees

1. Two men, *A* and *B,* are running for office. Each man gets *n* votes. In how many ways can the 2*n* votes be counted so that at no time is *A* behind *B?* (1 = vote for *A,* 0 = vote for *B.*)

2. Place a penny, a nickel, and a dime in a row. On the penny put a stack of *n* face-up playing cards with values in consecutive order from bottom to top. The cards are moved one at a time from the penny to the nickel or from the nickel to the dime. (No other moves are allowed.) By mixing these two types of moves, you will end, after 2*n* moves, with all the cards on the dime. Given *n* cards, how many different permutations can you achieve on the dime? (1 = move from penny to nickel, 0 = move from nickel to dime.)

3. An inebriated man leaves the door of a bar and staggers straight ahead. His steps are equal, but before each step, he has a random choice of going forward or backward. How many ways can he take 2*n* steps that will return him to the door? (1 = step forward, 0 = step back.)

This random walk can be given other forms. A king starts on the first row of a chessboard and moves one square forward or back along the file to end on the starting square after $2n$ moves. Draw a space-time diagram of the moves, with time measured along the horizontal base line. The zigzag path can be viewed as the profile of a mountain with peaks an integral number of miles high and a base length of $2n$ miles. The paths depict all mountain ranges of this type.

4.   An even number ($2n$) of soldiers, no two the same height, line up in two equal rows, A and B. How many ways can they do it so that from left to right in each row the heights are in ascending order and each soldier in row B is taller than his counterpart in row A? (Number the soldiers 1, 2, 3, . . . according to increasing height, and number the digits of the binary numbers from left to right. The 1 digits give the numbers for row A, and the 0 digits give the numbers for row B. The problem is easily modeled with playing cards.)

5.   Tickets are 50 cents, and $2n$ customers stand in a queue at the ticket window. Half of them have \$1 each and the others have 50 cents each. The cashier starts with no money. How many arrangements of the queue are possible with the proviso that the cashier always be able to make change? ($1 = 50$ cents, $0 = \$1$.)

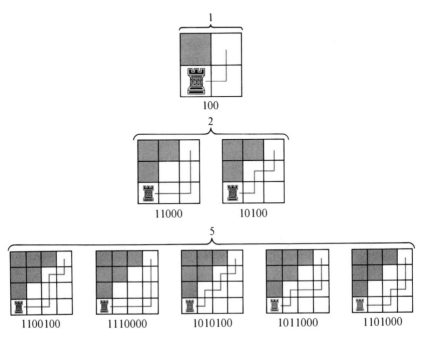

Figure 130   How Catalan numbers count a rook's paths

6. Hexaflexagons are curious toys made by folding straight or crooked strips of paper into a hexagonal structure that alters its "faces" when it is flexed. (They are described in the first chapter of my *Scientific American Book of Mathematical Puzzles & Diversions.* Simon & Schuster, 1959.) A regular hexaflexagon of a specified type passes through different states as it is flexed. The total number of states, for all varieties of regular hexaflexagons of *n* faces, is a Catalan number. For example, a hexahexaflexagon (six faces) can be made in three ways. The total number of states is the Catalan number 42.

If we ignore the states and ask in how many essentially different ways a regular hexaflexagon of *n* faces can be made, the answer is provided by a sequence that counts the triangulations of convex polygons when rotations and reflections are excluded. This remarkable sequence (No. 942 in Sloane's *Handbook*) is 1, 3, 4, 12, 27, 82, 228, 733, 2282, 7528, ...

In unpublished papers, Bernhart and other flexagation addicts describe ways of mapping the changes of states, as a flexagon of *n* faces is flexed, by tracing paths around the lines of a triangulated polygon of $n + 1$ sides.

7. An even number of people are seated around a circular table. Each extends one arm and they clasp hands in pairs, but in such a way that no pair of joined arms crosses another. Given the number of pairs, in how many ways can this be done? More precisely, place $2n$ spots in fixed positions on the circumference of a circle and then find all the ways they can be paired by drawing nonintersecting chords.

Can you find a simple geometric way to establish a one-to-one correspondence between this problem and any of the above problems?

A nonrecursive formula for the *n*th Catalan has different forms depending on how the positions of the Catalans are numbered. The formula is simplest if the sequence begins 1, 2, 5,. . . . In this numbering, the *n*th Catalan is

$$\frac{(2n)!}{n!(n + 1)!}$$

If the series begins 1, 1, 2, 5, . . . , it turns out that odd Catalan numbers greater than 1 appear at all positions, and only at those positions, that are powers of two. Thus the fourth, eighth, sixteenth, and so on Catalans are odd. This is only one of many unusual properties of the sequence that have been discovered.

A word of caution: When one works on combinatorial problems, it is easy to confuse the Catalan sequence with a closely related one: 1, 2, 5, 15, 52, 203, 877, ... As Gould points out in notes on his bibliography (which also includes a separate listing of references on the above series), when structures are complicated, it is easy to miss a fifteenth structure (when $n = 4$) and to be tricked into supposing you have encountered a Catalan sequence. The numbers are called Bell numbers after Eric Temple Bell, who published a lot about them. They count the partitions of n elements. For example, the number of rhyme schemes for a stanza of n lines is a Bell number. A quatrain has fifteen possible rhyme schemes. A 14-line sonnet, if convention is thrown to the winds, can have 190,899,322 (the fourteenth Bell) distinct rhyme schemes. But, you may object, who would write a sonnet with a rhyme scheme such as *aaaaaaaa aaaaaa*? Allowing a word to rhyme with itself, James Branch Cabell conceals just such a sonnet (each line ending with "love") in Chapter 14 of *Jurgen* (Grosset and Dunlap, 1919). I would guess it no accident that Cabell's 14-line poem starts as the fourteenth paragraph of Chapter 14.

Figure 131 shows how Bell numbers count the rhyme schemes for stanzas of one line through four lines. Lines that rhyme are joined by curves. Note that not until we get to quatrains does a pattern (No. 8) require an intersection. Joanne Growney, who worked this arrangement out in 1970 for her doctoral thesis, calls the schemes that do not force an intersection of curves "planar rhyme schemes." Bell numbers count all rhyme schemes. Catalan numbers are a sub-sequence that counts planar rhyme schemes.

The Bell sequence is No. 585 in Sloane's *Handbook*. But the Bells chime another story that we must postpone for a future book.

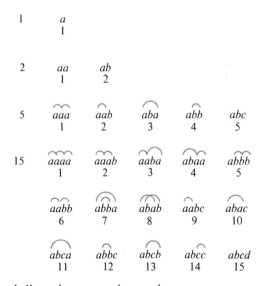

Figure 131 How bell numbers count rhyme schemes

ANSWERS

The problem was to show that if $2n$ spots on a circle are joined in pairs by nonintersecting chords, the number of ways of doing so are counted by the Catalan numbers. Figure 132 shows how Frank Bernhart establishes a one-to-one correspondence of the chord patterns with planted normal trees of $n + 1$ edges. It also gives their binary numbers. Since these trees are counted by Catalan numbers, as I explained, the same sequence counts the chord patterns. To make the diagrams easier to interpret, the small chords are shown curved.

Imagine that the circles and chords are elastic strings. Break each circle at the top and bend it into a straight line. The chord problem then becomes: $2n$ points on the line are paired in all possible ways by joining them with curves above the line that do not intersect. This is equivalent to finding all "planar rhyme schemes" for $2n$ lines that consist of n couplets, allowing coupled lines to be separated by other lines.

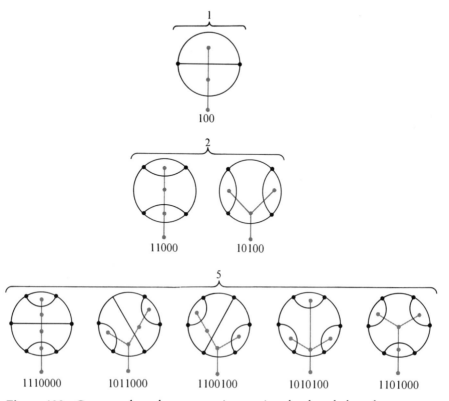

Figure 132 Correspondence between nonintersecting chords and planted trees

If the line is now closed by bringing its ends together *below,* we get an inside-out version of the original problem. It could represent a lake with $2n$ houses on the perimeter, paired in all possible ways by nonintersecting paths. (All paths joining a given pair of houses are assumed to be the same; think of each pair as being joined by an elastic string that can be lifted up, stretched or shrunk and replaced anywhere on the plane outside the lake.)

To obtain the binary numbers, imagine a worm at the bottom of each circle, inside the circle and facing west. As the worm crawls clockwise around the circle, it calls out 1 when it encounters a chord for the first time and 0 when it encounters a chord a second time. The procedure works in reverse. Given the binary number, the worm labels the spots 1 and 0. There will be only one way to join each 1 to a 0 without an intersection of chords.

ADDENDUM

My column on Catalan numbers produced so many letters telling me about other applications of the numbers, and other properties, that I can mention only a few of special interest.

Vern Hoggatt, Jr., who edited *The Fibonacci Quarterly,* explained how easily the Catalans can be found in Pascal's famous number triangle. Merely go down the center column (1, 2, 6, 20, 70, . . .) and from each number subtract the adjacent number (numbers on left and right of the central number are, of course, the same). Result: the Catalan sequence!

Paul Stockmeyer called my attention to Jack Levine's "Note on the Number of Pairs of Non-Intersecting Routes" (Scripta Mathematica 24, Winter 1959, pp. 335–338). Stockmeyer's colorful interpretation of Levine's result is to imagine two people at the same intersection of a square grid. Each simultaneously goes one step, randomly choosing to go either north or east. When their paths intersect they outline a polyomino. The Catalans count the number of distinct polyominoes that can be formed after each person has taken n unit steps. The theorem underlies many later papers, such as "A Catalan Triangle," by Louis Shapiro (*Discrete Mathematics* 14, 1976, pp. 83–90.)

Shapiro also sent me his paper, "A Short Proof of an Identity of Touchard's Concerning Catalan Numbers" (*The Journal of Combinatorial Theory,* 20, May 1976, pp. 375–376.) Here is Shapiro's interpretation of the identity. Put n points on a circle. Each point is either painted red, or green, or joined by a line to another point. The lines must not cross. Catalan numbers count the number of different patterns for each n.

My column on the related sequence of Bell numbers ran in *Scientific American* (May 1978) and will be included in a later book collection.

B I B L I O G R A P H Y

"Some Problems in Combinatorics." H. G. Forder, in *The Mathematical Gazette* 45, October 1961, pp. 199–201.

"A Note on Plane Trees." N. G. de Bruijn and B. J. M. Morselt, in *Journal of Combinatorial Theory* 2, January 1967, pp. 27–34.

"Prime and Prime Power Divisibility of Catalan Numbers." Ronald Alter and K. K. Kubota, in *Journal of Combinatorial Theory* 15, November 1973, pp. 243–256.

"Catalan Structures and Correspondences." Mike Kuchinski. Master's thesis (University of West Virginia), privately published by the author in Morgantown, West Virginia, 1977.

"The Computation of Catalan Numbers." Douglas Campbell, in *Mathematics Magazine* 57, September 1984, pp. 195–208.

TWENTY-ONE

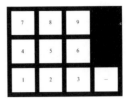

Fun with a Pocket Calculator

If you could climb into a time machine and go back to ancient Athens for a visit with Aristotle, what could you carry in your pocket that would most astonish him? I suggest it would be a pocket calculator. Its Arabic number system, its light-emitting diodes, its miniaturized circuitry isomorphic with Boolean logic (Aristotle, remember, invented formal logic), and above all, its computational speed and power would intrigue him more than any other small object I can think of.

The revolutionary consequences of these miraculous little gadgets are only beginning to be manifest. Among engineers and scientists the slide rule has already become as obsolete as the abacus. It is sad to think of the mathematicians of recent centuries who devoted years to the arduous calculation of logarithms and trigonometric functions. Today an engineer finds it takes less time to calculate such numbers all over again on a pocket machine than to look them up in a book or make a slide-rule approximation.

Among mathematics teachers, controversy over the "new math" has been replaced by controversy over how pocket calculators should be used in elementary education. Almost everyone agrees that, at the high school level and above, the machine will be an enormous boon. "It is unworthy of excellent men," wrote Leibniz (who invented a mechanical computer), "to lose hours like slaves in the labor of calculation." Freed from such drudgery, students surely will be more inclined to study the basic concepts and structures of mathematics. It is no credit to our education that when a mathematician discloses his profession to a stranger, he waits for the inevitable remark: "I can't even balance my checkbook." Would you tell a poet or a novelist about your spelling difficulties?

Teachers are arguing not about the ultimate value of the pocket calculator but about when it should be introduced. The consensus is that it should not be until a child has learned how to add, subtract, multiply, and divide on paper. After that, there seems to be no good reason for not allowing students to take a calculator to class or even to use one in examinations. In any case, the revolution is unstoppable. Already there is talk of having calculators built into desks, like old-fashioned inkwells.

The recreational-math buff who buys even one of the least expensive calculators will soon be wondering how he ever managed without it. Consider a cryptarithm such as

$$
\begin{array}{r}
\text{TH I S} \\
\text{I S} \\
\hline
\text{* * T O O} \\
\text{HARD *} \\
\hline
\text{* * * * * *}
\end{array}
$$

Few tasks could be more boring than solving this puzzle without a calculator. It is apparent that S must be 2, 3, 4, 7, 8, or 9 (otherwise S and O would not be distinct), and I cannot be 0, 1, or the same as S. One can easily determine by hand that IS must be 72, 57, 68, or 79. From here on, however, there are no good clues, and most people need a calculator to run through the possible values of T and H in a reasonable time. (We assume the usual conventions: Each letter stands for only one digit, different letters stand for different digits, asterisks represent any digit, and numbers do not begin with 0.)

There are so many other ways that pocket calculators stimulate interest in both serious and recreational mathematics that I can touch on only a few high spots. If the machine has a memory key that makes it possible to hold the partial sums of a converging infinite series, it can be of enormous value in guessing the limit before searching for a proof. For example, take this unfamiliar series in which the numerators are the odd numbers in sequence and the denominators form the doubling series:

$$1/1 + 3/2 + 5/4 + 7/8 + 9/16 + \cdots$$

After each division add the result to the previous sum. The partial sum, after 10 fractions have been added, is 5.95. . . . The series seems to be converging on 6. Is that the limit?

Try entering any number in the readout and then repeatedly pushing the square-root key. You will see the roots quickly converge on 1. Suppose that after each square-root extraction, you double the result before taking the next root. Is the limit 2? No, it is 4. Instead of doubling, each time multiply by m. The limit proves to be m^2. Generalize further by taking repeated nth roots (if your machine can do it) and multiplying each result by m. Can you write and prove the formula for the limit? (I am indebted to Don Morran for this problem.)

Several books have been written on competitive games between two or more players that make use of pocket calculators, but few of the games use the machine for more than rapid calculation. Lynn D. Yarbrough's "keyboard game" is a pleasant exception. It appeared in the special games and puzzles issue (January 1976) of the magazine *Creative Computing*.

The keyboard game begins with the first player entering any positive integer, say 100. The second player punches the subtract key, then any digit key on the three-by-three array (*see* Figure 133), and finally the equals key. The game continues with players alternately subtracting digits (0 excluded) until a player loses by activating the minus sign.

There is one condition that prevents this game from being trivial. On each turn, after the first subtraction, a player must choose a key adjacent (ortho-

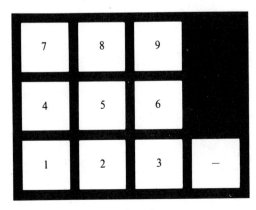

Figure 133   Three-by-three array for keyboard game

gonally or diagonally) to the key last pressed. Thus if 5 is played, the next player may subtract any digit except 5. If 4 is played, the next play is limited to 1, 2, 5, 7, and 8. If 3 is played, it is limited to 2, 5, 6, and so on.

Without this proviso, the first player can win easily by choosing a multiple of 10 for the initial number and then subtracting the right-hand digit of the readout whenever it is his turn. With the proviso, the game is fun to play. Better still, it has a surprising solution. It turns out that the second player (the one who makes the first subtraction) can always win regardless of the initial number. If the beginning number is greater than 15, he wins by punching either 1 or 3 (it makes no difference which) until the total is 13 or less, then playing carefully thereafter. Interested readers can consult Yarbrough's article for the complete strategy and for variants of the game when 0 is allowed.

One of the more curious recreational uses for the pocket calculator is as a device for performing magic tricks, most of them of the ESP variety. Here is a stunt that children find particularly amusing. Ask a child to enter 98765432 and to divide by 8. He or she will be mildly surprised by the result: 12345679. The digits are in sequence except for 8, the divisor, which has mysteriously vanished.

Ask the child to name his favorite digit. Suppose he says 4. You immediately say: "Very good. Multiply the number on display [12345679] by 36." Now he is really surprised, because the number he gets is 44444444 (or nine 4's if the readout can accommodate that many). The multiplier you give is always the product of 9 and the named digit. The working is easy to understand: $111111111/9 = 12345679$. Since 9 times 12345679 is 111111111, a multiplier of $9n$ (where n is a digit) is sure to give a row of n's.

Dividing a row of 1's by integers other than 9 until the quotient has no fractional remainder produces other "magic numbers." For example: $111111/7 = 15873$. Multiplying 15873 by $7n$, where n is a digit, produces a row of n's. Again: $111111/33 = 3367$. Multiply 3367 by $33n$ and you get a row of n's.

A trick I call the Arabian Nights Mystery, because it is based on the properties of 1001, begins by asking someone to think of any three-digit number, ABC. Tell him to repeat the number to make $ABCABC$ and enter the six-digit number in the calculator. While he is doing it, stand with your back turned so that you cannot see what is going on.

"I'm beginning to get some vibes," you say, "and they tell me that your number is exactly divisible by the unlucky number 13. Please divide by 13 and tell me if I'm right."

Your companion makes the division. Sure enough, there is no remainder.

"It's strange," you continue, "but my clairvoyant powers tell me that the number now on display is exactly divisible by the lucky number 11." He makes the division. You are right again.

"Now I have a strong impression," you continue, "that the number on display is exactly divisible by the still luckier number 7." This proves to be the case.

Tell your companion to take a good look at the readout. It is *ABC,* the number he first thought of.

The trick cannot fail. Multiplying any three-digit number *ABC* by 1001 obviously produces *ABCABC.* Because the prime factors of 1001 are 13, 11, and 7, dividing *ABCABC* by those three numbers must result in *ABC.*

One of the oldest and still one of the best of all number-guessing tricks is particularly interesting because it introduces a celebrated theorem that goes back to a book on arithmetic believed to have been written in the fourth or fifth century by a Chinese mathematician and poet named Sun-tsu (or Sun-tse). One of the book's problems is to find the smallest natural number (positive integer) such that when it is divided by 3, 5, and 7, the remainders are respectively 2, 3, and 2. Sun-tsu supplies the answer, 23. He also gives, in verse, a general rule that he calls the *t'ai-yen* (great generalization) for solving the problem.

Guessing a number from 1 through 315, by using the divisors 5, 7, and 9, appears in the medieval arithmetic (1202) of Leonardo Fibonacci, the Italian mathematician for whom Fibonacci numbers are named. Tricks of this form were popular throughout the Middle Ages and the Renaissance. They can be presented with any number of divisors, provided that they are relatively prime (have no common divisor) and that the chosen number is not greater than the product of all the divisors. (The divisors themselves need not be prime. For example, they could be 3 and 4, with the chosen number ranging from 1 through 12.) Since a pocket calculator makes for fast computing with divisors larger than digits, let us see how the trick operates with 7, 11, and 13, our lucky and unlucky numbers. Their product is 1001, so we can safely ask someone to think of any number from 1 through 1000.

The trick is most effective if no one knows you are using a calculator. Hand someone a pencil and paper, then cross the room and sit with your back toward the spectators. Surreptitiously take the calculator out of your pocket and use it on your lap.

Ask your companion to think of any number not greater than 1000, divide it by 7, and tell you the remainder. He repeats this procedure twice more, dividing the original number by 11, telling you the remainder, then dividing the original number by 13 and telling the remainder. How do you calculate the selected number?

Let a, b, and c be the three remainders. The chosen number is the remainder after making the following calculation:

$$\frac{715a + 364b + 924c}{1001}$$

The three coefficients should be memorized or written on a small strip of paper pasted on the calculator. These are the simple steps you follow:

1. While your companion is dividing his chosen number by 7, enter 715. Multiply it by the announced remainder. If your calculator has no memory key, you will have to jot down the product and also the next two products so that you can add them later. If your machine has a memory that allows chain addition, enter the product in the memory.

2. While your companion is dividing his chosen number by 11, enter 364. Multiply by the announced remainder and add it to the preceding result.

3. While he does his final division by 13, enter 924. Multiply by the announced remainder and add to the preceding sum. The number now in your readout is equal to your companion's chosen number modulo 1001. If it is less than 1001, it is the number. If it is greater, reduce it to the chosen number by repeatedly subtracting 1001 until the number on display goes below 1001.

How was the formula derived? The derivation is best explained with an example, so let us use Sun-tsu's simpler version. The divisors are $a = 3$, $b = 5$, $c = 7$, and the chosen number must be no greater than 105.

The coefficient of a is the lowest multiple of bc that is one more than a multiple of a. There are rules for finding this coefficient, but when the divisors are small, as they are in this case, it is easy to get the number by inspection. Simply go up the multiples of bc (35, 70, 105, . . .) until you come to a multiple that has a remainder of 1 when it is divided by 3. The multiple is 70.

The other two coefficients have similar forms. The second coefficient is the lowest multiple of ac that is one more than a multiple of b. It is 21. The third coefficient is the lowest multiple of ab that is one more than a multiple of c. It is 15. We can now write the formula:

$$\frac{70a + 21b + 15c}{105}$$

The number below the line is abc. This ancient version of the trick is the one most popular today with mathemagicians. It accommodates a chosen number from 1 through 100, and the formula is simple enough so that with practice the calculations can be done in the head. The mental steps can be simplified by replacing $70a$ with $-35a$, since -35 is equal to 70 modulo 105. This procedure also keeps the total lower, so that fewer subtractions of 105 are needed to reach the chosen number.

The remarkable theorem behind such tricks is called the Chinese Remainder Theorem in honor of Sun-tsu. In general, it states: Given a set of relatively prime natural numbers greater than 1 $(d_1, d_2, \ldots, d_n)$ and a corresponding set of natural numbers $(r_1, r_2, \ldots, r_n)$, there is a unique number modulo x (where x is the product of all the d numbers) such that, when it is divided by d_i, the remainder is $r_i \pmod{d_i}$ for every value of i.

This is one of the most valuable theorems of congruence arithmetic. It serves not only for proving deeper theorems but also for answering many practical questions. Early astronomers and astrologers employed it for solving problems concerning solar, lunar, and planetary cycles. Oystein Ore's *Number Theory and Its History* (McGraw-Hill, 1948) gives several applications of the Chinese Remainder Theorem to ancient puzzles as well as to a systematic procedure for the splicing of telephone cables. In 1967 Elwyn R. Berlekamp used the theorem in developing a fast algorithm for the computer factoring of polynomials. [A useful reference is Section 4.6.2 of *Seminumerical Algorithms*, Volume 2 of *The Art of Computer Programming* (Addison-Wesley, 1969) by Donald E. Knuth. Kurt Gödel made use of the theorem in his famous undecidability proof.]

Another remarkable number theorem, which goes back to Fibonacci, underlies a prediction trick recently proposed by Francis T. Miles. Write the numerals 1, 6, and 8 on a piece of paper and turn it face down without letting anyone see what you wrote. Someone now uses the calculator to generate three "random" digits by the following method. He writes down any number he likes and writes below it any second number. Below that, he puts the sum of the two numbers. The third number (the sum) is then added to the second number to get a fourth. This procedure is continued (each time by adding the last sum to the preceding number, using the calculator when the numbers get large) until the list has twenty numbers. Tell your companion to divide the last number by the preceding one, or vice versa if he prefers, and to take note of the first three digits of the decimal fraction. They are almost certain to be the three digits you predicted.

The trick works because in a generalized Fibonacci sequence, which is what the spectator is generating, the ratio between adjacent terms approaches the golden ratio, 1.618033 . . . , as a limit. It does not matter which number is divided by which, because the reciprocal of the golden ratio is .618033. . . . A magician I know likes to predict four decimal digits by placing four playing cards face down on the table. After turning over a six, an ace, and an eight, he looks crestfallen, as if he had missed on the fourth digit, since there is presumably no zero card. Then he turns the fourth card to reveal a blank face. If the fourth digit is not zero, he says, "As you see, I don't take chances."

No article on play with pocket calculators would be complete without mentioning the recent proliferation of whimsies exploiting the fact that most of the readout digits resemble letters when they are viewed upside down. Magic magazines were the first to publish such jokes, but they became widely known only after they were discussed in *Time* (June 24, 1974, pp. 56–58). The procedure is to ask a question, then give a series of computations to produce a result that spells the answer when it is inverted.

The earliest of these upside-downers seems to have involved questions about who won a certain skirmish between Arabs and Israelis. After punching in the relevant information, one got the answer: 71077345. Inverted it spells SHELLOIL. Knuth devised the most mathematically interesting story line: 337 Arabs and 337 Israelis battle over a square property 8424 meters on a side. Naturally we sum the squares of 337 and 8424. Is there another way to obtain 71077345 as the sum of two squares? Yes, there is just one additional way: $5,324^2 + 6,537^2$. Several books dealing entirely with such inversions have been published.

My off-color contribution to this useless pastime has appeared only in magic periodicals. What do Congress and belly dancers have in common? Multiply the prime number 2417 by the number of months in the year, divide by the number of letters in "Congress," then multiply by the number of letters in "George Washington." Turn the machine around to read the answer. For greater precision, add 1.0956 to the number on display, then subtract .1776.

The number 1776, by the way, in addition to being a famous date, has the following curious property. Select any positive digit N. Punch the calculator's N key three times to put NNN in the readout. Multiply by 16, then divide by N. The result is always 1776.

ANSWERS

The solution to the cryptarithm is

$$
\begin{array}{r}
4379 \\
79 \\
\hline
39411 \\
30653 \\
\hline
345941
\end{array}
$$

The problem appears in Joseph S. Madachy's *Mathematics on Vacation* (Scribner's. 1966). Other questions are answered as follows.

To prove that the series

$$1/1 + 3/2 + 5/4 + 7/8 + \cdots$$

converges on 6, first halve each term to obtain

$$1/2 + 3/4 + 5/8 + 7/16 + \cdots$$

Subtract this from the original series:

$$
\begin{array}{l}
1/1 + 3/2 + 5/4 + 7/8 + 9/16 + \cdots \\
\quad -\ 1/2 + 3/4 + 5/8 + 7/16 + \cdots \\
\hline
1\ \ +\ \ 1\ \ +1/2 + 1/4 + 1/8\ \ + \cdots
\end{array}
$$

The sequence at the right of the first term has a sum of 2, so the entire series must have a sum of 3. Since this series is half of the original series, the original has a sum of 6.

If we begin with any natural number and then repeatedly take the nth root and multiply it by m, the limit L is

$$m^{n/(n-1)}$$

Don Morran's proof is as follows. Assume that $mL^{1/n}$ approaches a nonzero limit L. At the limit $mL^{1/n} = L$; $m^n L = L^n$; $L^n - m^n L = O$; $L(L^{n-1} - m^n) = O$; $L = m^{n/(n-1)}$.

ADDENDUM

Since I wrote this column, magicians interested in mathematics have invented hundreds of new tricks, games, and stunts that use a pocket calculator, as well as more elaborate tricks requiring special computer programs. I will add here only one clever trick, invented by Karl Fulves, that anyone can perform.

Turn your back and ask someone to select a row, column, or either of the two main diagonals on a calculator's square of nine digit keys. He puts the three selected digits, in any order, into the readout. Ask him to select another row, column, or diagonal. He then multiplies the number on display by a number consisting of the three newly-selected digits, again in any order.

With your back still turned, ask him to choose any nonzero digit in the product, then to call out to you (in any order) the remaining digits. You correctly tell him the chosen digit.

The working of the trick depends on the fact that every row, column, and main diagonal contains three digits that add to a multiple of 3. Any permutation of those digits will, of course, leave the sum unchanged. This assures that any three-digit number formed by those digits will be a multiple of 3. The product of two such triplets is certain to be a multiple of 9, with digits that add to a multiple of 9, and this also will be the case regardless of how the digits are permuted.

As the spectator calls out the digits, add them in your head, "casting out nines" as you go. This is done as follows. Whenever a sum is more than 9, add the digits of the sum to obtain a single digit. After the last digit is called, subtract it from 9. The remainder will be the spectator's chosen digit, with one exception. If the remainder is 0, the selected digit is 9.

If you repeat the trick, Fulves suggests the following variation. Instead of crossing out a digit, ask the spectator to think of any digit (except 0) and add it to the number in the readout. He then calls out, in any order, the digits in the result, and you tell him the digit he selected.

B   I   B   L   I   O   G   R   A   P   H   Y

Games Calculators Play. Wallace Judd. Dymax, 1974; Warner Books, 1976.

The Pocket Calculator Games Book. Edwin Schlossberg and John Brockman. Vol. 1, William Morrow, 1975. Vol. 2, William Morrow, 1977.

The Calculating Book: Fun and Games with Your Pocket Calculator. James T. Rogers. Random House, 1975.

Calcu/Letter. Dan Steinbrocker. Pyramid Books, 1975.

Compute-the-Words Puzzles. Marvin D. Frost. Privately published, 1975.

Games with a Pocket Calculator. Sivasailam Thiagarajan and Harold Stolovitch. Dymax, 1976.

The Conjurer's Calculator. John C. Sherwood. Mickey Hades, 1976.

"The Magic Calculator." Martin Gardner, in *The New York Times Magazine,* January 18, 1976, p. 71.

TWENTY-TWO

Tree-Plant Problems

Your aid I want, nine trees to plant
In rows just half a score;
And let there be in each row three.
Solve this: I ask no more.

—JOHN JACKSON

Look at the back of a dollar bill and you will see above the eagle on the Great Seal of the United States thirteen five-pointed stars—symbols of the thirteen original states—arranged as a six-pointed star. The pattern shows that 13 is the first nontrivial (larger than 1) figurate number of a type called "star numbers." (See Chapter 2.)

There are, of course, a multitude of other ways to arrange thirteen points on the plane to meet the demands of aesthetic symmetry or the provisos of recreational mathematicians. In a moment, we shall consider two unsolved problems, far from trivial, that concern ways of arranging thirteen points. But first let us take a look at how the Colonists arranged the thirteen bright stars on their earliest flags.

According to popular legend, the first stars-and-stripes flag was sewn by Betsy Ross, who based it on a rough drawing supplied by George Washington. She is said to have displayed her handiwork to Washington and others in May or June of 1776 at a house somewhere on Arch Street in Philadelphia. To show how she made the pattern for her stars, she is said to have folded a sheet of paper and then, with one snip of her scissors, to have cut out a perfect pentagram. The thirteen stars, so the story goes, were arranged on a blue field in a circle to imitate King Arthur's Round Table. It is the first design in Figure 134. In 1976 our Bicentennial year, the flag was on a 13-cent stamp. It appears in such famous paintings as "Washington Crossing the Delaware" by Archibald M. Willard, "The Spirit of '76," and the picture by Henry Mosler that used to be in many schoolbooks showing Betsy's nimble fingers at work on Old Glory.

Alas, the story has been totally discredited. Its sole source was Betsy's grandson, who said he heard it when he was eleven. Not a single flag with thirteen stars in a circle has survived, and there is no evidence that such a flag existed in Revolutionary times. Many historians doubt that a stars-and-stripes flag of *any* design flew during a single sea or land battle of the Revolution.

The second design in the illustration shows how the stars are arranged on a flag that John Hulbert, captain of a company of minutemen from Long Island, is said to have flown in 1775. This arrangement too is not supported by any evidence. The truth is that no one knows who designed the first stars-and-stripes flag, and almost nothing is reliably known about the flag's earliest history.

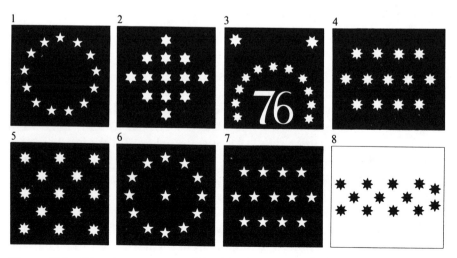

Figure 134 Alleged arrangements of the thirteen stars on early flags of the United States

We do know that on June 14, 1777, the Second Continental Congress resolved "that the flag of the United States be thirteen stripes, alternate red and white; that the union be thirteen stars, white in a blue field, representing a new constellation." There is not a word about the number of star points, how the stars are to be arranged, or which color of stripe should outnumber the other. As a result, from 1777 until 1795 there were wild variations in flag designs. Some flags even violated the Congressional order by having red and blue stripes or red, white, and blue stripes; some had blue stars on a white background.

The stars had five, six, seven, or eight points and were arranged in all kinds of ways. Number 3 in the illustration is on a flag said to have been used by the Bennington militia in 1777. Numbers 4 and 5 appear in paintings by a Dutch artist on flags alleged to have been flown by John Paul Jones on ships in 1779. Number 5, with its field stretched horizontally, was perhaps the most common pattern on flags from 1777 to 1795. (Note that it is Number 2 rotated 45 degrees.) Number 6 is supposed to have been on the flag of a Maryland regiment and number 7 on a flag flown in Boston, both around 1781. Number 8 is credited to the North Carolina militia of 1781.

In 1794, after Vermont and Kentucky had become states, Congress decided to add two more stars and two more stripes, arranging the stars as is shown in Figure 135. One congressman thought this was "a trifling business which ought not to engross the attention of the House." Another called it "a consummate specimen of frivolity." But the bill was passed, and the flag became official the following year. This was the flag that Francis Scott Key is alleged to have seen flying over Fort McHenry during the War of 1812 and that inspired him to write "The Star-spangled Banner."

In 1817, after five more states had joined the union, Congress decided to go back to thirteen stripes and to have a new star added each time a state was admitted. A flag with twenty stars in a four-by-five rectangle became official in 1818. From then until 1913 there were twenty-four changes of the flag as stars were added. The stars were usually five-pointed and in staggered rows, although many other arrangements were popular, including star-shaped patterns. After the admission of New Mexico and Arizona in 1912 the flag was stable for forty-six years, its forty-eight stars arranged in a six-by-eight rectangle. In 1959, to accommodate a new star for Alaska, President Eisenhower ordered a seven-by-seven field, the rows staggered with odd rows aligned at the left. When Hawaii became a state later in 1959, he ordered the field changed to its present form of fifty stars in staggered rows of six and five.

There are many puzzles based on the arrangement of n spots on a field, but the oldest and most popular are known as "tree plant" problems. They have

Figure 135 The flag of 1794, with fifteen stripes and fifteen stars

that name because in early puzzle books they were usually presented with a story about a farmer who wishes to plant a certain number of trees in an orchard so that the pattern of trees will have r straight rows of exactly k trees in each row. The puzzles were made difficult by maximizing the number of rows. Surprisingly, the general problem of determining the largest number of rows, given n and k, is nowhere near solved even when k is 3 or 4.

"These tree-planting puzzles," wrote Henry Ernest Dudeney, England's greatest puzzle expert, in *Amusements in Mathematics* (Dover, 1958), "have always been a matter of great perplexity. They are real 'puzzles' in the truest sense of the word, because nobody has yet succeeded in finding a direct and certain way of solving them. They demand the exercise of sagacity, ingenuity and patience, and what we call 'luck' is also sometimes of service. Perhaps some day a genius will discover the key to the whole mystery."

When k is 2, the problem is trivial. If n points are arranged so that no three are in line, every pair forms a row of two. When k is 3, the problem not only becomes interesting but also is related to such mathematical topics as balanced-block designs, Kirkman-Steiner triples, finite geometries, Weierstrass elliptic functions, cubic curves, projective planes, error-correcting codes, and many other aspects of significant mathematics. The latest and most definitive paper on the topic is "The Orchard Problem," by Stefan A. Burr, Branko Grünbaum and N. J. A. Sloane. What follows is taken mainly from that paper.

The first nontrivial reference on tree-plant problems is a book called *Rational Amusement for Winter Evenings,* by John Jackson, published in London in 1821. According to Dudeney, who owned a copy, it contains ten such puzzles. The mathematician J. J. Sylvester worked continually on the general problem from the late 1860s until his death in 1897. (The temperamental Sylvester had a stormy career. He was denied a degree at Cambridge because of his Jewish faith but obtained one at Trinity College of the University of Dublin. He was a professor at the University of Virginia for three months until an altercation with a student led to his resignation. At Johns Hopkins he founded *American Journal of Mathematics.* One of his books is *The Laws of Verse* [he was fond of writing poetry], and for many years in England he was a barrister.)

Maximum solutions for three-in-a-row tree plants, from three through eleven points, are shown in Figure 136. Note that not until *n* equals 9 does the maximum number of rows exceed *n*. It is easy to get eight rows with nine points (a three-by-three square array does it), but adding two more rows is a bit tricky. In his book Jackson introduced the problem with the quatrain that is this chapter's epigraph. The solution, derived from a famous theorem in projective geometry called the Pappus theorem, has been credited to Isaac Newton.

The eleven-point pattern is given by Dudeney (Problem 213 of his *Amusements*) as a military puzzle. In lecturing on tree-plant problems, Sloane has simplified Dudeney's narrative line by describing a World War I battlefield on which eleven Turks were surrounded by sixteen Russians. Each Russian fired once, and each bullet passed through exactly three Turkish heads. How were the Turks standing? The remarkable solution — eleven points in sixteen rows of three each — is said by Dudeney to have been constructed about 1897 by the Reverend Mr. Wilkinson. (Does any reader know who he was?) Sloane tells me that this is the only practical application of the orchard problem he knows, although he once invented a fictional "Haltwhistle triode" of *n* pins, which, because of unexplained capacitance effects, have to be arranged in rows of three.

Twelve points will make nineteen rows. This result was announced, apparently for the first time, by R. H. Macmillan in a 1946 note to *The Mathematical Gazette* and is proved maximal in the Burr-Grünbaum-Sloane paper. Figure 137 (left) shows a way of drawing the pattern symmetrically by placing three points and one line at infinity. This pattern can be projected to give a standard solution, but it is difficult to show on a small sheet of paper. Imagine the pattern viewed in perspective with the eye below and to the right. Each of the three sets of four parallel lines (labeled with *a*'s, *b*'s, and *c*'s) will converge, the three meeting points lying on the horizon to form the nineteenth row.

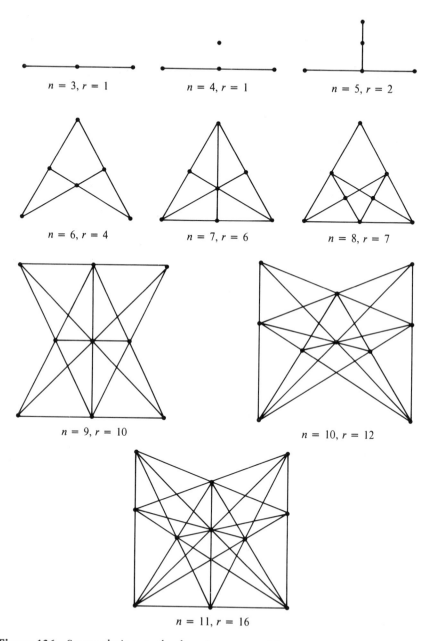

Figure 136   Some solutions to the three-in-a-row problem

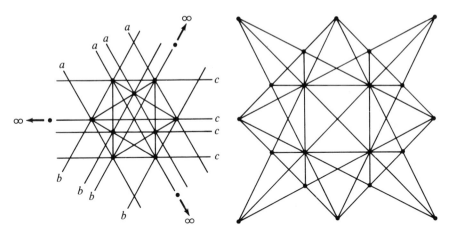

Figure 137 Twelve points in nineteen three-point rows (left) and Sam Loyd's solution for rows of four

Only one other three-in-a-row case has been solved, that of $n = 16$. Burr, Grünbaum, and Sloane prove the maximum number of rows to be 37. Thus the lowest unsolved case is $n = 13$. The best-known result, twenty-two lines, is shown in Figure 138. One point is at infinity. If the pattern is viewed in perspective from the left, the six parallel horizontal lines, of two points each, will converge on the thirteenth point. In other words, the pattern can be projected to give a standard solution, but it is difficult to show it except on a large sheet. The best-known results for n equals 14 through 20 are 26, 31, 37, 40, 46, 52, and 57.

When the number of points demanded for each row rises to four, the problem becomes more difficult. As in the case of $k = 3$, maximums have been established through twelve points, with thirteen as the lowest unsolved case. Examples of best patterns from four through twelve points are shown in Figure 139. They are taken from Grünbaum's "New Views on Some Old Questions of Combinatorial Geometry," a paper he gave at the International Colloquium on Combinatorial Theories in Rome in 1973. The best known results for $n = 13$ through 20 are 9, 10, 12, 15, 15, 18, 19, and 21.

The case of $n = 10$ has many topologically distinct solutions (see Chapter 2 of my *Mathematical Carnival*, Knopf, 1975), providing Dudeney and his American rival, Sam Loyd, with more than a dozen puzzles. When n equals 16, the best-known result (fifteen rows) is an elegant pattern of three nested pentagons surrounding a central point (*see* Figure 140). In *The Canterbury Puzzles and Other Curious Problems* (Dover, 1958), where this arrangement solves Problem 21, Dudeney admits he cannot prove it but says he has a "strong pious

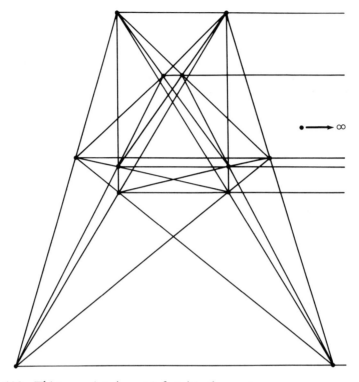

Figure 138   Thirteen points (one at infinity) in three-point rows

opinion" that fifteen rows is maximal. It is surprising and infuriating that no one has done better with seventeen points than this same pentagonal pattern with the seventeenth point added as a total irrelevancy.

The pentagonal pattern appears in the photograph of a blossom of *Hoya carnosa,* a member of the milkweed family. Imagine the sides of the outer petals extended to points as shown in Figure 141. The three largest sets of pentagonally placed vertexes, together with the flower's center, beautifully give the fifteen rows of four each in the best-known solution for twenty points.

The number of rows of four begin to exceed the number of points when n equals 21 (for which 23 is believed maximal), but when n equals 18, 19, or 20, rows equal to n are possible. Figure 142 shows how Grünbaum gets nineteen rows with nineteen points. The case of $n = 20$ is answered in two of Loyd's puzzle books with eighteen rows (*see* Figure 137, right). In about 1945 Macmillan found a simple, symmetrical way to make twenty rows with twenty points. It was later rediscovered by Grünbaum, who was unaware at the time of Macmillan's unpublished results. Can readers construct it?

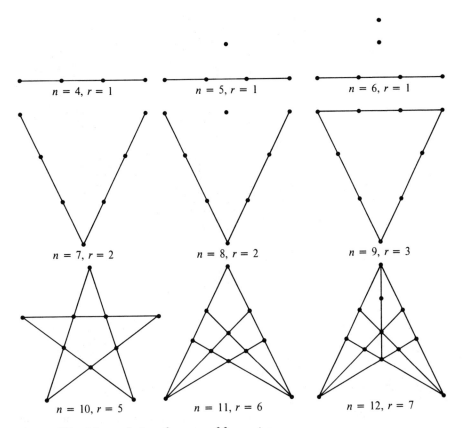

Figure 139 More solutions for rows of four points

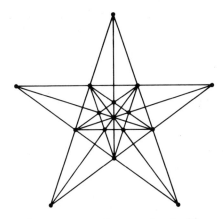

Figure 140 Dudeney's flower pattern: sixteen points in fifteen rows

Figure 141 A blossom of *Hoya carnosa,* a milkweed, in a pattern of points

When *n* equals 13, no one has done better than the nine rows shown in Figure 143. Dudeney gives this as the solution to Problem 149 of his *Modern Puzzles and How to Solve Them,* reprinted as Problem 435 in *536 Puzzles & Curious Problems* (Scribner's, 1967), a collection of his puzzles.

Very little work has been published on rows of five or more points. According to Grünbaum, the best results known for *k* = 5, points five through twenty, are 1, 1, 1, 1, 2, 2, 2, 3, 3, 4, 6, 6, 7, 9, 10, and 11. I do not know how many of them have been proved maximal. A set is known of thirty-five symmetrically arranged points that determine thirty-six rows of five. Grünbaum conjectures that no smaller example exists in which the number of rows of five exceeds the number of points. I know of no work that has been done on the extension of tree-plant problems to spaces of three dimensions or more.

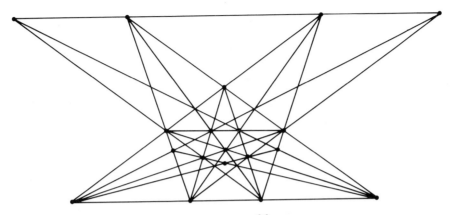

Figure 142   Nineteen points in nineteen rows of four

Burr, Grünbaum, and Sloane conjecture that, with the four exceptions of *n* equals 7, 11, 16, and 19, the formula for the maximum number of rows of three, given *n* points, is

$$1 + \left[\frac{n(n-3)}{6} \right]$$

The brackets indicate rounding down to the nearest integer. If the conjecture is correct, no one can do better with thirteen stars than twenty-two rows. For rows of more than three points there are not even good conjectures.

ANSWERS

Figure 144 shows how to solve the problem of placing twenty points in twenty rows of four points each. Note how the petals of the milkweed flower

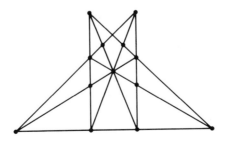

Figure 143   Thirteen points in nine rows of four

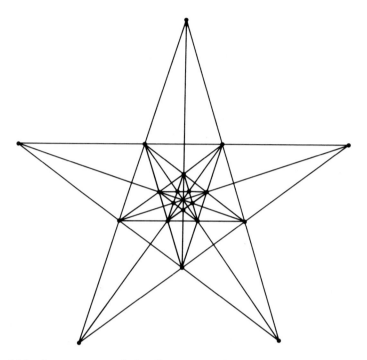

Figure 144 A twenty-row solution for twenty points

almost provide this solution. Unfortunately, the little pentagram close to the flower's center is not quite large enough.

ADDENDUM

The biggest surprise in my mail on this chapter, when it ran as a column, were two letters that improved on the pattern shown in Figure 144. Ton van Teeseling of Amsterdam and Douglas McClean of Cape Town each went one better by finding solutions with twenty-one rows! Teeseling's pattern requires three points at infinity to get it on a page. McClean's solution, which I ran in a later column, requires no points at infinity.

The pattern I published was not left-right symmetric. Leonard Lopow was the first of many readers who observed that McClean's pattern could be made symmetric by shifting one point. Lopow found a second solution, more compact and also symmetric, that is shown in Figure 145. There is no proof that twenty-one is maximal. "I hesitate to offer my 'pious opinion,'" Lopow

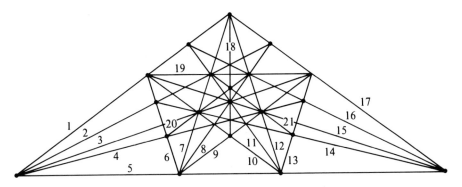

Figure 145 A twenty-one row solution for twenty points

commented, "but I'll take off twenty-two hats to the guy or gal who does better than twenty-one rows."

T. H. Wilcocks wrote from England to pose an interesting problem involving tree plants. Given a specific task, find the smallest rectangular checkerboard on which the solution can be given by placing checkers on the cells. For example, the 7-by-7 board is the smallest that will accommodate ten checkers in five straight lines of four each. It provides a pleasant puzzle that was published and answered in *Games* magazine (July 1982).

Tree-plant patterns are often useful in solving what combinatorialists call tournament and committee problems. Mention also should be made of two areas of recreational mathematics that are based on the patterns:

Can consecutive integers, starting with 1, be placed on the points of a given pattern so that all its rows have a constant sum? If so, find all solutions; if not, prove impossibility.

Can interesting ticktacktoe-like games be devised in which two players alternately put a counter on a point until a player wins by filling a row with his counters, or loses by being the first who is forced to complete such a row? Thomas H. O'Beirne seems to have been the first to explore games of this sort. See his article in *New Scientist* cited in the Bibliography.

B I B L I O G R A P H Y

"An Old Problem." R. H. Macmillan, in *The Mathematical Gazette* 30, 1946, p. 109.

"New Boards for Old Games." T. H. O'Beirne, in *New Scientist* 269, January 11, 1962, pp. 98–99.

Arrangements and Spreads. Branko Grünbaum. American Mathematical Society, 1972.

"The Orchard Problem." Stefan A. Burr, Branko Grünbaum, and N. J. A. Sloane, in *Geometriae Dedicata* 2, 1974, pp. 397–424.

"New Views on Some Old Questions of Combinatorial Geometry." Branko Grünbaum, in *Teorie Combinatorie* 1, 1976, pp. 451–468.

"Planting Trees." Stefan A. Burr, in *The Mathematical Gardner,* David Klarner, ed. Prindle, Weber, and Schmidt, 1981.

"The Tree Planting Problem on a Sphere." Stephen J. Ruberg, in *Mathematics Magazine* 53, January 1981, pp. 41–42.

Adams, John Quincy, 31
Ali, Muhammad, 67–68
Allen, Woody, 5
Ammann, Robert, 238
Anderson, Poul, 13
Andrews, W. S., 219, 221
Appel, Kenneth, 134
Aquinas, Saint Thomas, 190
Archibald, R. C., 121
Archimedes, 189
Aristotle, 189, 267
Arrow, Kenneth J., 56, 57
Asimov, Isaac, 2

Bach, C. P. E., 89
Bach, Johann Sebastian, 87
Ball, W. W. Rouse, 39, 146, 148, 223
Banach, Stefan, 153
Banner, Randolph W., 115
Barnard, F. A. P., 223
Barnes, Frank, 238
Baron, G., 79
Beck, Anatole, 158
Beeler, Michael, 42, 216
Beidler, John, 162
Beiler, Albert H., 17–18
Bell, A. W., 183
Bell, Eric Temple, 4, 263
Bellamy, Edward, 5
Benford, G. A., 4
Benham, Philip S., 210
Benson, William, 223, 224
Berent, Paul, 244
Berlekamp, Elwyn, 160, 273
Berloquin, Pierre, 111
Bernhart, Frank, 257, 262, 264
Bessy, Bernard Frénicle de, 215
Beurdeley, Michel, 101
Bickel, P. J., 247
Biot, Jean Baptiste, 91
Bird, David, 181, 182, 183
Bishop, Gregory J., 182, 183
Black, Duncan, 57
Black, Max, 242
Block, Ned, 9–10
Blyth, Colin R., 247, 248–249
Boas, R. P., Jr., 119
Book, D. L., 4

Borges, Jorge Luis, 7, 85–86
Borgmann, Dmitri, 227
Borromini, Francesco, 106
Botermans, Jack, 51
Bradbury, Ray, 7
Braithwaite, R. B., 242
Brezhnev, Leonid, 128
Brillhart, John, 136
Brown, Fredric, 3, 8–9
Brown, G. Spencer, 94
Brown, Spencer D., 210
Brualdi, Richard A., 233–235
Bruijn, Nicolaas G. de, 228–229
Buller, A. H. Reginald, 3–4
Burke, Bernard F., 121
Burr, Stefan A., 280, 281, 283, 287

Cabell, James Branch, 112, 120, 263
Cahill, B. J. S., 197, 198
Calegari, Antonio, 90
Calvin, John, 189
Cammann, Schuyler, 215
Candy, Albert L., 216
Carnap, Rudolf, 243, 245, 249
Carroll, Lewis, 31, 57, 200, 207,
    210–211
Catalan, Eugene Charles, 256
Cayley, Arthur, 257
Chernick, Paul, 249
Chess, David, 24
Chopin, Fréderic, 92
Chuan-Chih Hsiung, 42
Clarke, Andrew L., 182
Clinton, J., 91
Cohen, Jacques, 47
Condorcet, Marie Jean, Marquis de, 57
Confucius, 30
Conway, John Horton, 52, 65, 66, 67,
    76–77, 159, 178–179, 180, 182,
    233, 235, 237
Corey, Irwin, 13
Coxeter, H. S. M., 126, 146, 148, 206,
    223
Crapper, Thomas, 131, 137
Crowe, Donald W., 139, 141, 143, 144,
    145–146
Császár, Ákos, 140
Cummings, Alexander, 130

Dali, Salvador, 103
Daniels, David R., 7
Dante Alighieri, 190
De Morgan, Augustus, 228
Deutsch, E. S., 36, 46, 47
DeWitt, Bryce, 10, 11–12
Dickson, Leonard E., 15
Dingle, Herbert, 136
Dodgson, Charles Lutwidge. *See* Carroll, Lewis
Doré, Gustav, 31
Dowling, L., 91
Dudeney, Henry Ernest, 30, 31–35, 215, 216, 280, 281, 283–284, 285, 286
Dunsany, Edward Hohn, Lord, 7–8

Einstein, Albert, 2
Eisenhower, Dwight D., 279
Elffers, Joost, 52
Escher, M. C., 164, 171, 184, 185
Euler, Leonhard, 16, 17, 23, 254–255, 256
Evans, Ronald J., 158–159
Everett, Hugh, III, 8, 10–11, 12

Felver, Richard I., 121
Fermant, Pierre, 23, 24
Feynman, Richard P., 2, 4
Fibonacci, Leonardo, 271, 273
Fisch, Forest N., 209
Fischer, Bobby, 128
Fisher, Irving, 198, 199, 201
Fodor, Nandor, 138
Fontaine, Alan, 107
Forder, H. G., 256
Foregger, Thomas H., 233–235
Foreman, George, 67–68
Forward, Robert, 13
Franklin, Benjamin, 214
Frazier, Joe, 67–68
Fremlin, David, 161
Fu Tsiang Wang, 42
Fuller, R. Buckminster, 197–198
Fulves, Karl, 52–53, 161, 275, 276

Gamow, George, 2, 136
Ganrei-Ken (pseud.), 51
Gardner, Jim, 138
Geller, Uri, 133
Gernsback, Hugo, 137
Gershwin, George, 91

Ghyka, Matila, 23
Giessen, William C., 210
Gilbert, Edgar N., 194, 195, 196, 197
Giuliani, Baron J., 91
Godel, Kurt, 2, 4, 12, 56, 273
Godwin, Joscelyn, 88
Golomb, Solomon W., 53, 151, 177–178, 182
Good, I. J., 136
Goodman, Nelson, 244
Gosper, William, 222
Gould, Henry W., 254, 262–263
Graham, L. A., 120
Graham, Neill, 10, 11
Green, Roger L., 210
Greenblatt, Richard, 136
Grinstead, Charles M., 24
Growney, Joanne, 263
Guibas, L. J., 67
Grünbaum, Branko, 280, 281, 283, 284, 286, 287
Guy, Richard K., 160

Hadamard, Jacques, 147
Hadsell, Alan, 81
Hagmann, Mark J., 138
Haken, Wolfgang, 134
Halmos, Paul, 58, 113
Hamilton, John F., Jr., 249–250
Hammel, E. A., 247
Hao Wang, 182
Harington, Sir John, 130
Harvis, John, 20, 23, 24–25, 118, 123–124, 182
Harrison, Richard Edes, 198, 199, 201
Hatch, Evelyn M., 211
Haydn, Joseph, 89
Hayes, Kenneth C., Jr., 47
Hein, Piet, 158, 202
Hempel, Carl G., 243–244
Hendricks, John, 224
Herstein, I. N., 120
Hikoe Enomoto, 124
Hindin, Harvey J., 24
Hoffmann, Professor (pseud.), 51
Hogan, Randolph, 95
Hoggatt, Vern, Jr., 265
Holbein, Hans, 98, 99
Holvenstot, Clyde E., 121
Hooper, W., 95
Hulbert, John, 278
Hume, David, 243

Hurst, Bryce, 67
Husserl, G., 208

Ittelson, W. H., 105

Jackson, Brad, 123
Jackson, John, 281
Jacobs, Frank, 95
Jacoby, Oswald, 223–224
James, Richard E., III, 171–172, 173, 174
Jeans, Sir James, 86
Jefferson, Thomas, 108
Jeffrey, Richard C., 250
Johnson, Samuel, 31
Johnson, Selmer M., 74
Johnson, Wesley, 24
Johnson-Laird, Philip, 82
Jones, John Paul, 279
Joplin, Scott, 94

Kaplansky, Irving, 113, 120
Kaprekar, D. R., 115–117, 122
Katona, G., 231
Keeler, Donald, 121
Keller, David M., 120
Kemeny, John G., 243
Kershner, Richard Brandon, 164, 165, 166–171, 172
Key, Francis Scott, 279
Kilpatrick, F. P., 105
Kircher, Athanasius, 87, 88F
Kirnberger, Johann Philipp, 87
Kissinger, Henry, 128
Kiyoshi Ishihata, 124
Klarner, David A., 228, 231, 235, 236F, 238
Knuth, Donald E., 60, 76, 273, 274
Koch, Tom, 95
Kraft, Dean, 133
Krippner, Stanley, 138
Kuhn, Thomas S., 243
Kulagina, Nivel, 133

Landsberg, P. T., 136
Langford, C. Dudley, 78
Langman, Harry, 224
Laszlo, Alexander, 89
Lehmer, D. H., 72, 76, 136
Leibniz, Baron Gottfried, 267–268
Lenard, Andrew, 60
Levine, Eugene, 79

Levine, Jack, 265
Levy, Steven, 124
Lewthwaite, G. W., 154–156, 161
Lindgren, Åke, 42
Lindgren, Harry, 40
Lindon, J. A., 9, 23
Lopow, Leonard, 288–289
Loyd, Sam, 28, 29–30, 31, 32, 35, 36F, 98, 100, 108, 283, 284
Lukasiewicz, Jan, 257
Lull, Ramón, 87
Luther, Martin, 189
Lynning, Ejvind, 47
Lyons, P. Howard, 80

McClean, Douglas, 288
McCulloch, Warren S., 58
McGregor, William, 123, 126, 135
McKay, John, 23
MacMahon, P. A., 183–184
Macmillan, R. H., 281, 284
Madachy, Joseph S., 275
Madan, Falconer, 210
Mandelbrot, Benoit, 95
Marinoni, Augusto, 137
Mason, J. H., 150
Mather, Michael, 238
Mauldon, James G., 223, 224
Mazur, Stanislaw, 153
Meally, Victor, 23, 223
Mencken, H. L., 29, 137
Mendeléev, Dmitri Ivanovich, 12
Mendelsohn, N. S., 113, 120
Mercator, Gerhardus, 192
Micheli, Romano, 88
Miles, Francis T., 273
Mill, John Stuart, 87, 242, 250
Miller, O. M., 198
Mines, Sam, 3
Minkowski, Hermann, 2, 5
Mitchell, Edward Page, 1
Mizler, Lorenz Christoph, 87
Mollweide, Karl B., 192
Monge, Gaspard, 121
Moon, J. W., 58—60
Mordell, Louis J., 20
Morran, Don, 269, 275
Moser, Leo, 58–60
Moskowitz, Sam, 1, 7
Mosler, Henry, 278
Mozart, Wolfgang Amadeus, 89–90, 94

Murray, Sir James, 30
Myers, Richard Lewis, Jr., 219, 221–223

Napoleon Bonaparte, 31
Nash, John F., 158
Nelson, Norman N., 209
Newcomb, W. A., 4
Newton, Isaac, 281
Niceron, Jean François, 98, 103, 106
Nye, Bill, 93

O'Beirne, Thomas H., 89, 90, 93, 289
O'Connell, J. W., 247
Odlyzko, A. M., 67
Ogilvy, C. Stanley, 120–121
Ore, Oystein, 273
Ostrander, Sheila, 138

Papaikonomou, Angeloo, 81
Parsons, Denys, 94
Pascal, Blaise, 265
Pavalita, Robert, 138
Paz y Remolar, Ramón, 137
Peirce, Charles Sanders, 196–197
Pell, John, 23
Penney, Walter, 60
Penrose, L. S., 185
Penrose, Roger, 184–185
Pepys, Samuel, 87
Philpott, Wade, 182
Piel, Anthony, 68
Piel, Gerard, 198
Pierce, J. R., 91
Pinkerton, Richard C., 91–92
Planck, C., 224
Plato, 189
Poe, Edgar Allan, 31, 32
Popper, Karl, 243
Post, J. B., 200
Priestley, J. B., 7
Putnam, Hilary, 5–6, 8
Pythagoras, 189

Ramanujan, Srinivasa, 126, 136
Randi, James, 134
Ransom, Tom, 78, 80, 82, 83, 205,
    206F, 208–209
Ravielli, Anthony, 137
Read, Ronald C., 28, 30, 31, 35, 36,
    40–42, 43, 46, 48, 50
Rebertus, Dennis, 161
Reichenbach, Hans, 5, 6, 8, 242

Reinhardt, K., 165, 166, 167, 168
Reiss, Richard, 51–52
Reyburn, Wallace, 131, 137
Reti, Ladislao, 129
Rice, Marjorie, 174, 175
Rindler, W., 136
Ringel, Gerhard, 123
Robinson, Raphael M., 185
Room, Thomas G., 146
Rorschach, H. E., 119
Roselle, D. P., 79
Ross, Betsy, 278
Rosser, J. Barkley, 223
Roth, Norman K., 134
Rucker, Rudy, 67
Russell, Bertrand, 242

Sachs, David, 67
Sackson, Sidney, 53
Samuelson, Paul A., 56
Satoru Kawai, 124
Sawade, Kanue, 151
Schattschneider, Doris, 174
Schillinger, Joseph, 91, 92
Schmucker, Kurt, 149, 150
Schor, J. Paul, 88F
Schroeder, Lynn, 138
Schroeppel, Richard, 216, 218, 219,
    220, 221
Schuyt, Michael, 52
Schwenk, Allen J., 122
Segner, Johann Andreas von, 255
Segre, Beniamino, 135
Sei Shonagon, 51
Shannon, Claude E., 228
Shapiro, Louis, 265
Shaw, A., 91
Shaw, R., 136
Shear, David B., 121
Shigeo Takagai, 51
Shoji Sadao, 199
Silver, Roland, 81
Silverman, David L., 66, 68, 78, 80, 81,
    161
Simpson, E. H., 247
Sin Hitotumatu, 24
Singmaster, David, 231
Sleator, Daniel, 121
Sloane, N. J. A., 23, 253–254, 262,
    263, 280, 281, 283, 287
Slocum, Jerry, 51
Sluizer, Allan L., 42

Smith, Cedric A. B., 160
Snyder, Jay, 82
Spencer, Herbert, 121
Stadler, Maximilian, 89
Staib, John H., 221
Standen, Maud, 207
Stein, Rolf, 174
Steinhaus, Hugo, 74
Stewart, Bonnie M., 149
Stockmeyer, Paul, 265
Stolarsky, Kenneth B., 122
Stromquist, Walter, 120
Sun-tsu (Sun-tse), 271, 272, 273
Sweet, John Edson, 114, 120
Sylvester, J. J., 281
Szász, D., 231

Takeuchi, Dr. I., 42
Tannenbaum, Herbert, 101
Taylor, Edwin F., 136
Teed, Cyrus Reed, 200
Teeseling, Ton van, 288
Thomasson, T. C., Jr., 79
Thompson, D'Arcy Wentworth, 107
Tipler, Frank, 12–13
Titzling, Otto, 137
Tobler, Waldo R., 196, 202
Tromelin, Count de, 138
Trotter, H. F., 74
Twain, Mark, 5, 7, 8, 93

Ulam, Stanislaw, 58, 153, 160, 162

Van Delft, Pieter, 51
Vandersteel, Stoddad, 81
Van Gulik, 31
Van Note, Peter, 30, 52

Vero, Radu, 121
Villa-Lobos, Heitor, 91, 92F
Vinci, Leonardo de, 98, 129, 130, 137
Vonnegut, Kurt, 2
Voss, Richard, 95
Vout, Colin, 156–158

Wagner, Richard, 93
Walker, Richard, 121
Walker, Robert J., 223
Wallis, John, 23
Warshaw, Stephen I., 210
Washington, George, 98
Wason, Peter, 82
Wayne, Alan, 206
Weinrich, James D., 82–83
Weizsäcker, 12
Wells, H. G., 1–2, 3, 5
Werner, Johann, 195
Wheeler, John A., 8, 12, 136
Whitford, E. E., 23
Wigner, 12
Wilcocks, T. H., 289
Wilcox, L. R., 172
Wilf, Herbert S., 81–82
Wilkinson, Rev. Mr., 281
Willard, Archibald M., 278
Williams, Sidney H., 210
Wilquin, Denys, 111–112
Winder, Robert O., 158–159
Winkel, M., 91
Wolk, Barry, 62–63, 66
Wrench, J. W., Jr., 119
Wright, Lawrence, 131

Yarbrough, Lynn D., 269, 270